George Hebard Paine

The New Roadmaster's Assistant

A Manual of Reference for Those Having to do With the Permanent Way of

American Railroads

George Hebard Paine

The New Roadmaster's Assistant
A Manual of Reference for Those Having to do With the Permanent Way of American Railroads

ISBN/EAN: 9783744678407

Printed in Europe, USA, Canada, Australia, Japan

Cover: Foto ©berggeist007 / pixelio.de

More available books at **www.hansebooks.com**

THE NEW

ROADMASTER'S ASSISTANT

A MANUAL OF REFERENCE FOR THOSE HAVING TO DO WITH

THE PERMANENT WAY OF AMERICAN RAILROADS

BY

GEORGE HEBARD PAINE

This volume is the successor to the Roadmaster's Assistant written by
William S. Huntington in 1871 and revised by
Charles Latimer in 1877

PUBLISHED BY
THE RAILROAD GAZETTE, 32 PARK PLACE NEW YORK
1898

To my Father, CHARLES PAINE, as a slight evidence of my admiration, gratitude and affection, this book is dedicated.

TABLE OF CONTENTS.

CHAPTER.	PAGE.
1. General Remarks,	1
2. Organization and Methods of Work,	9
3. Fences, Highway Crossings and Platforms,	19
4. Miscellaneous Fixtures and Station Grounds,	31
5. Water Supply,	47
6. Drainage,	55
7. Culverts, Trestles and Bridge Floors,	67
8. Ballast,	75
9. Cross Ties,	95
10. Rails and Fastenings,	107
11. Track Work,	127
12. Tools,	147
13. Frogs, Switches and Switch Stands,	165
14. Emergencies and Train Signals,	193
15. Fixed Signals,	201
16. Rules and Tables,	229
Index,	255

PREFACE.

In his preface to the "Road-Master's Assistant and Section-Master's Guide," Mr. William S. Huntington explained that his wish was to make a "practical book for practical men." How well he succeeded is shown by the great demand for it from men of all ranks in railroad service, while to those just entering the maintenance of way department it has been a guide and counsellor.

To the labor begun in 1871 by Mr. Huntington, the revision made by Mr. Charles Latimer in 1877 brought the scope of the work down to his period; nor did he, in the fullness of his knowledge and experience, depart from the rule that Mr. Huntington had adopted.

It has been a pleasant but difficult task to follow in the footsteps of these men; to be clear, concise and comprehensive but without dogmatism, or a too great insistence on his own opinion; to be practically useful and theoretically correct has been the object of the present author.

Nearly all maintenance of way knowledge is based upon the experience of predecessors; few can truthfully say, "I was the father of that idea," since the present practice is the outcome of more than a half-century of labor and thought. But of all this travail a large proportion has been mistaken, the proof of which lies before us in the scrap-heap.

Nevertheless, our mistakes of to-day beget the successes of to-morrow; therefore it behooves us of the present to regard our views with diffidence, in the certain assurance that many notions which we consider sound will on trial prove wrong, and much which we now condemn is to be the practice of the future.

The drawings in this book are original and their correctness and simplicity are largely due to Mr. Arthur Tartas. The author is indebted to Mr. James D. Hawks, President of the Detroit & Mackinac Railway Company, and to Mr. Augustus Torrey, Chief Engineer of the Michigan Central Railroad Company, for advice and criticism.

G. H. P.

March, 1898.

CHAPTER I.

GENERAL REMARKS.

An experience of something like sixty years *Construc-* in railroad construction has demonstrated the *tion.* fact that the preparation of a piece of track for the passage of trains can be done best, cheapest and most quickly by well-organized gangs of men, each of them under the command of one who is experienced in his particular branch of the work. For this reason, a modern railroad is usually built and perhaps ballasted by contractors and is turned over to the maintenance-of-way department with the tracks and switches laid.

The first duties which fall upon a road-mas- *Finishing* ter, after receiving charge of a newly built *up.* line, are the not very agreeable ones of finishing up. Contractors are quite likely to leave their work in an unfinished condition, and the permanent officers of a railroad then find that the cuts and banks are not sufficiently sloped; that the track needs re-lining and re-surfacing; the ties re-spacing; and that the ditches, if any have ever been dug, are nearly, if not quite, filled up.

After having seen that the track is in a safe *Opening* condition and that loose rocks and rotten trees *streams.* are not likely to fall upon it, the roadmaster should at once make an inspection of all waterways. The beds of the streams should be cleared of all loose material, both above and below, as well as underneath the openings;

this should be particularly looked after in the case of wooden structures where the danger of fire is added to the liability of washouts.

Cleaning culverts. Culverts are frequently built too small to accommodate even ordinary high water; therefore if they are permitted to remain choked with weeds and driftwood they become doubly dangerous.

Cleaning ditches. Because good track cannot exist with bad drainage, prompt attention should be paid to the condition of the ditches, which should be opened as soon as possible and in such a way as to permit of the quickest passage of the water from the ballast to some regular water course.

Frogs and switches. Frogs and switches should be rigidly inspected and, although mistakes in their design or construction cannot often be remedied after they are once in place, it should be ascertained that guard rails are properly located and braced; that the throw of the switches is correct; that their points are sufficiently protected by the curve in the main rail and that the frog points are in line with the main track.

Trees near track. Trees are often left standing on or near the right-of-way in such a position, if they should be blown over, as to endanger passing trains. These must be felled in short order, although their removal may be put off to some less busy time. In some states railroads are permitted to condemn trees outside the right-of-way (if a price cannot be agreed upon with the owner), and it is important for trackmen to know of this wherever the right exists. A simple and sufficiently accurate way of determining whether or not a tree is at a safe distance from the track, if the base of the tree is neither much above or much below the level of the track, is as follows: Let one man hold a track-gage vertically, resting it on top of the rail nearest the tree.

Let another man place his eye close to the opposite rail and sight over the upper end of the track-gage. If the sight line clears the top of the tree, the tree is at a safe distance — otherwise, not.

Generally speaking, the only vegetation which should be permitted on the right-of-way of a railroad is grass and this should be encouraged in every way, for it is of the greatest use in preventing sliding in cuts and on other slopes and much improves the appearance of a line as seen from a passing train. *Encouraging grass.*

Section-foremen should be authorized to employ a limited number of men to assist them in times of threatened danger from floods, landslides, etc. If the privilege is abused it may easily be taken away or the foreman replaced by a more judicious one; since it is better to spend a few dollars at the right time than that the whole traffic of a railroad should be stopped indefinitely for want of a little extra help. *Extra men.*

Snow storms, especially, should be promptly dealt with, and as it devolves upon the maintenance-of-way department to keep the switches and platforms clean, roadmasters and section-foremen should be prepared at all times to meet any storms with a sufficient force to perform their work in a satisfactory manner. *Snow storms.*

The track should be completely walked over by the roadmaster at least three times, and better still four times, a year, in the company of each section-foreman, for the purpose of making a general comparison of progress and planning the work for the future. These walks should take place at the opening of spring, early in July, the middle of September and at the beginning of winter. Between times, each section should be attended to as occasion requires, while the fact should be borne in mind *Track inspection.*

that the best way to inspect a piece of track is to do it on foot.

Reports. Although the accounts and reports which are required of roadmasters and section-foremen are not difficult, they are apt to cause considerable trouble to the men and annoyance at the department head-quarters. The only *easy* way is to have them ready when they are due and to have them right, since delaying the performance of a duty makes it harder to fulfill and increases the likelihood of error. The section time-books should be made up every night after the quitting time, and, while the matter is still fresh in the mind, all of the charges to different kinds of work should be made. The tie, rail and material reports of the roadmaster should be kept in a form which will enable him to fill them up and send them in at the end of the month, without having to spend several days in the office at a time when he should be out on the road among his men.

Use and abuse of hand-cars. The use of the section hand-cars by a roadmaster should be resorted to only on rare occasions and under the most urgent necessity. Although it is, for the roadmaster, an extremely pleasant and convenient method of getting from place to place, it is expensive for the railroad company, and, what is worse, leads to lax habits on the part of the men. For ordinary touring a velocipede should be used.

The hand-car should never be used except in charge of the foreman himself, or someone in whom he has confidence, and when on the track should be the object of care and watchfulness. Where the trains are frequent, on a crooked road, during a fog, or at night, the car should be protected by a flag, and the fact that a hand-car has been struck by an engine should be regarded as presumption of criminal carelessness on the part of the foreman. The cars should not be "taken off" on highway crossings in such a way as to block them, but frequent places, formed of old rail, ties or earth, should be provided for the purpose.

It should never be taken from the house without the following equipment: two red flags, one green flag, six torpedoes, a well-sharpened mattock, an oil can, a monkey wrench, a spike maul, a track chisel and a claw bar. Other tools will be found convenient, but with those named, the track can be protected and nearly any kind of small repairs can be made. *Hand-car equipment.*

Let it be always remembered that men cannot work without food. When they are kept out late at night in the cold or wet it puts them in good humor and gives them new strength, to supply them with sandwiches or bread and butter and hot coffee. For this purpose the work train caboose, as well as the wrecking car, should be provided with boilers for making coffee. The master who sees to the comfort of his men will, other things being equal, have more influence over them and get more out of them than the one who treats them with indifference. *Comfort of men.*

Reliable men should never be dismissed except for cause or to comply with a general order for reduction. The frequent discharge of employes for trivial reasons tends to breed dissatisfaction and uncertainty in the minds of the men who should be made to feel confident of keeping their positions during good behavior. On the other hand, a man who is discharged for a good and just cause should not be re-instated. Admonition should be tried, if it is possible, with men whose work is at all satisfactory, before the final act of discharge; but, whatever course is followed, it should be *Discipline.*

made clear to everyone that some notice will be taken of any careless or wanton breaking of the rules.

Knowledge of details. It is of the first importance that a roadmaster should know his road and his men, and it is more important that he should know the bad places and unworthy men than to know the good ones. Good things do not require so much watching as bad ones.

Emergency material. The roadmaster should also know the exact location of all material under his charge; and this knowledge is absolutely necessary at times when it is important that a large amount of material of a certain kind shall be delivered at a certain point at the earliest possible moment. Spare material, for the same reason, should be stored at convenient points and so placed as to permit of its being loaded easily and rapidly on short notice. If the roadmaster is prudent, he will always have some timber, rails, a few switches and some frogs on hand for sudden emergencies, no matter how poor the railroad company may be.

Intoxicants forbidden. The use of intoxicants should be absolutely prohibited during working hours. Men known to frequent saloons do not belong on a railroad, and for many reasons should not be employed there. The example they set is bad, and it cannot be foretold when someone in the humblest position may be required to perform work of immense importance, as for instance flagging a train which is in danger. At such a time a man must be in the full possession of his faculties, and if he drinks, he cannot be relied upon.

Competition. A spirit of competition and emulation once aroused among the men will prove a valuable help, and for this purpose tours of inspection, at stated intervals, over all the sections and shared in by all the foremen, should be made. They should be asked to criticise each other's work freely, and discussions as to the best way of accomplishing various things should be encouraged. This interchange of ideas will not only add greatly to the general stock of information but it will let each man see how far his work is advanced in comparison with that of the others.

At all times there should be held in line for promotion a number of bright, active young men who may be called upon to act as substitutes or to take the places of men whom it is desired to discharge. They may be familiarized with the use of authority by employing them as extra foremen during the summer, as track-walkers and upon detached service during the winter. The ability to enforce an order or inaugurate a reform will frequently depend upon this particular, since men are often retained in their positions for the sole reason that there are none to supplant them who will certainly do better. It must also be remembered that frequent small promotions have a better effect than a single considerable one; therefore in making a change it is well to see if two or three cannot be benefited instead of simply the one who is directly interested.

Men for promotion.

A section-foreman's place is with his men, whom he should not leave if it can be avoided. The roadmaster's place is everywhere. He should ride over his division continually; on the rear end and on the locomotives of passenger trains, on way freights and on a velocipede hand-car, while occasional trips should be made at night to see that the switch lamps are burning properly and that the track-walkers are attending to their duties. No man should undertake the duties of a roadmaster who will not cheerfully give himself up to the requirements of the work. He should be available at any hour of the day or night, and for this rea-

Attention to business.

Attention to business. son his whereabouts should always be known either at the telegraph office or at his home, and any serious damage to the main track should have his immediate personal attention. In short, his habits, life and language should be an example to his men in order that he may consistently correct any failures on their part. Inasmuch as he occupies one of the most responsible and onerous positions on the road, he should attempt to perform his duties with credit to himself or else earn his living in some other way.

CHAPTER II.

Organization and Methods of Work.

The number of men necessary to properly maintain a railroad is determined by such varying conditions that it is impossible to lay down any general rule which is applicable to every case. The amount and quality of the ballast, the condition and weight of the rail, the character and amount of the traffic, the climate; all these tend to affect the ease with which a piece of track may be kept up. On a well-ballasted, double-track railroad, equipped with good ties and heavy steel rails, having sections five miles long, five men and a foreman (exclusive of watchmen and track-walkers) for eight months beginning with April 1st, and three men and a foreman for the other four months of the year should be able to keep the road bed and track in first-class shape. This estimate is intended to cover only the routine work of a section with perhaps a little grading for a new side-track occasionally added. As an example of what is believed to be a remarkable economy in maintenance, it is stated that the Michigan Central Railroad employs but three men and a foreman in summer and two men and a foreman in winter on sections five miles long. This obtains on both single and double track since it is found that the increased train movement on single track fully compensates for the greater length of double-track sections. The Detroit & Mackinac Railway employs two men and a foreman in summer and one man

Number of men.

and a foreman in winter on eight to ten-mile sections of single-track road. Most of the tamping is shovel-tamping in sand and the traffic of course is light.

Length of sections. The sections, except those embracing large yards, should be as nearly as possible of equal length in order that a comparison may be made of the work performed by the different gangs. On main lines, sections should not exceed five miles in length, while on branch lines they may be seven or eight miles long, but seldom more because of the loss of time in going over them.

Extra help. As nearly as possible each foreman should perform all of the work of ordinary repairs on his own section. The practice of transferring one gang to help in the regular work on another section is not commonly a good one, since each foreman should be held responsible for, and capable of performing, his own duties, to which end he should be encouraged in every way. If the amount of work to be done is too great for the regular force, its number should be increased, but any help or interference from foreign gangs is apt to arouse the resentment of an ambitious man or to encourage in a lazy man a certain shiftless feeling of satisfaction.

Floating gangs. During the working months, a floating gang in charge of some bright young man, who is on the list for promotion to a regular section, will be found of great use in such matters as cutting new ditches, sloping rough banks, building new fences, etc., etc. These gangs are easily moved from place to place, may have their own hand-cars and tools, and for such work as they can do are useful and far more economical than a work train. If necessary (and it is frequently advisable) the foreman of the section where these men happen to be at work, may be permitted to guarantee their board and deduct it from their wages on pay-day.

The work train, although absolutely necessary for some purposes, is in many cases an expensive luxury. Since it has no rights beyond those given by special orders, it must keep out of the way of all regular trains and, if the road be a busy one, it is apt to become a loafing place for the men, because it must of necessity spend a large amount of time in running from one point to another or in waiting for orders at some place where there is no work to be done. To secure good results the roadmaster should watch its movements closely and work sufficient to keep the hands employed should be laid out some time in advance and distributed over the division in such a way as to provide constant employment for the men. *[Work trains.]*

The foreman of the work train should be bright, active and pushing, intent upon keeping his train in motion, always on the lookout for something to occupy his men and well acquainted with all the details of track work. The train should be provided with tools of all kinds in order that, no matter what service the men may be called upon to perform, they will have something to do it with. On long divisions and in some other cases the men must sleep and eat on the train. When this is so, it is a simple matter to have a man run the commissariat, charging an agreed price for each man per day. The railroad company can then collect his pay from the wages of the men and can also make the contract an object to the boarding boss by shipping his materials to him free and by furnishing the sleeping, dining and cooking cars.

Valuable help may be got out of the way freights, if properly handled, which is usually required of the work train. The distribution *[Use of way-freights.]*

of cross ties, small amounts of rail, ballast, building stone and other articles which need not delay the train long, may with a little foresight be thus accomplished at comparatively slight cost. Such work as heavy ballasting or ditching in long cuts requires an extra train, but by the use of an unloading plow, the presence of laborers on the train while the material is being handled and dumped may usually be dispensed with.

Combining gangs. A common method of concentrating labor is the massing together of men from a number of sections in order that a large amount of work may be accomplished in a short time. This plan is expensive but is applicable upon poor roads where the permission to employ extra men is not often granted.

Routine work. The routine work upon a railroad should be performed in a regular manner and in pursuance of well-considered plans. Certain days, as well as certain seasons, should be set apart for certain kinds of work. On Monday morning the section should be carefully inspected by the foreman for the purpose of remedying any defects which shall have developed during Sunday, and as much as is necessary of Saturday afternoon should be devoted to cleaning up. At this time the scrap, which in the meantime should have been thrown into small piles by the track-walker, must be collected and taken to the tool-house, the wrought and cast iron being thrown into separate bins. Unsightly objects should be disposed of and the line should be left in a neat and orderly condition.

Rainy days. There is always some work to be done around a section-house in the way of fitting handles, sharpening tools, slight repairs to the hand-car or house itself, which may be performed on rainy days when the men are waiting for it to clear up.

Watchmen. The watching of track is a most important feature of maintenance-of-way work, and is more likely to be neglected from a false sense of economy than it is to be overdone. In a properly organized department the slope watchmen will be furnished with the ordinary tools for tamping, renewing ties, ditching, etc., and will be expected to use them pretty steadily except in bad weather, when their attention should be directed to patrolling and looking for obstructions.

Track-walkers. Track-walkers, however, should not be required to make any but the most incidental repairs, such as knocking in a spike, tightening bolts, or some little thing which will not distract their attention or detain them long from their inspection. They should be required to look out for burning fences or fires which threaten, or appear likely to threaten, any property on or near the right-of-way. They should be particularly on the look out for broken rails, switches, frogs, etc. The position of track-walker is one of great responsibility and should be filled by a man of judgment, sobriety and experience in railroad work; one who is familiar with the book of rules, well acquainted with the time card, and able to immediately recognize a dangerous condition of the track or road bed. The track-walker is in fact the eye of the foreman. Many railroads do not employ track-walkers, and when this is so the section-foreman must be required to patrol his track every morning on a hand-car.

Foremen as laborers. It is a mooted point as to whether or not foremen should themselves work with their gangs, and a good deal is to be said on both sides of the question. In small gangs of three or four it is certain that the foreman should be able to perform a considerable amount of actual labor in addition to directing his men, but where there are eight or ten men to keep busy it is doubtful if anything is gained by distracting his attention from their movements.

Division of labor. As in the division of labor between the sections, it is advisable to divide the labor of a gang into several equal parts. Some men will always shirk unless there is a certain means of comparing their work with that of others who are not lazy.

Residence. The residence of foremen will be largely determined by circumstances, but if possible they should make their homes at stations where there are night telegraph offices. Roadmasters should live at some place from which they can easily reach all points lying within their jurisdiction.

Winter work. The railroad year, in most sections of this country, may be divided into two parts, the first of which extends from early in December until late in March. During this time as little as possible actual track work should be undertaken during the winter season. A little shimming, spiking and perhaps the renewal of switches and frogs where the ties are in good surface and do not require shifting, is about all that should be attempted and even these only for the purpose of keeping the track absolutely safe for the passage of trains. Many other useful things can be done, at odd times when the weather permits, such as ditching, repairing road crossings, fences and platforms, sloping banks, taking down loose rock, etc., which will greatly facilitate the work on the section. While cold weather lasts labor is cheap and plenty and advantage may be taken of this fact during moderate periods. The very last of the winter season is the best time to "renew rail." Since the ties must often be re-spaced in order that the rail may be properly supported at the joints, the most urgent part of this re-spacing must be done as soon as the frost is out of the ballast and may be finished when the new ties are put in.

The period from April to December should be devoted to the real work of the section. Lining and surfacing, although of great importance, stand second to ditching, which should begin as soon as the frost is out of the ground. If this fact were better understood much valuable time would be saved, for it is probable that any lining or surfacing of track before the ballast is thoroughly drained must be done over again in a very short time, and when trackmen are required or permitted to "putter," that is, rush from one point to another, picking up joints which are a little low, they will not have much time to do those things which most tend to preserve their track in good condition. There are only a few secrets not generally known about track work, and one of them is not to allow the section gang to "putter." *Summer work.*

After the ditching has been completed and the track made fairly smooth, the work of putting in ties should begin in earnest. For many reasons this part of the work should be pushed to as early a finish as possible, for, if it is carried on in a desultory manner, cold weather frequently arrives to find the work incompleted. No paying work can be done on track until the ties are in, and no good track can be secured unless ties are put in early in the season. The desirability of having all ties for the season's renewals on hand and distributed before the beginning of spring is therefore apparent. *Tie renewing.*

Weeds, shrubs and underbrush should be exterminated, and to this end should be cut at some specified time, which time varies in dif- *Cutting weeds.*

Cutting weeds. ferent parts of the United States. The object is of course to kill the weeds while they are small and before they ripen. In many of the states laws have been passed compelling the owners of land to cut the Canada thistles before a certain time in the summer under a penalty for neglect. The terms of this law, as it relates to their locality, should be known to every roadmaster and section-foreman. From this time on the force must be employed in the general work of the section, such as lining and surfacing, deepening water courses, laying drains, sodding banks and ballasting. About the middle of September a second cutting of the weeds will be found necessary, and this should be followed by a careful alinement and surfacing of the track and a general preparation of the road for the severities of winter.

Organized effort. Too much emphasis cannot be placed upon the necessity for organized effort in the maintenance-of-way department. There is a proper season for each different class of work and a regard for the old proverb which says "a place for everything and everything in its place" will secure as good results on a railroad as elsewhere.

Thorough work. It should be unnecessary to insist on so plain a fact as the importance of thorough work, but there is a too common idea that a piece of track which looks as if it were good is good "to stay." Hurried tamping may make track appear well when it is first put up, but a few trains passing over it cause it to become as bad as before, while the men are kept running from one low joint to another, losing time on the way, doing the same thing over and over again, when a little more effort in the first place would have resulted in a permanent job.

Order and neatness are of the first import- **Neatness.**
ance, not so much in themselves, as in what
they indicate. Although a foreman may have
a dirty car-house and good track, it is more
probable that ungathered scrap and other evi-
dences of carelessness will be accompanied by
loose bolts and badly tamped ties.

Reports should be required of all fires, with **Locomo-**
a statement of the numbers of the locomotives **tive**
which are believed to have caused them. Since **sparks.**
bad nettings are usually the cause, the locomotives which require repairs in that particular will be detected.

One of the most destructive agencies to a **Hollow**
railroad track are locomotive tires with hollow **tires.**
treads. On most railroads these tires are
turned down before they have been worn to a
dangerous depth, but occasionally a motive
power department is found which does not regard this fault so seriously as it should. To
correct it the roadmaster should occasionally
try the wheels of the locomotives by means of
a pocket template and enter a protest whenever a wheel is found with a hollow tire
deep enough to injure the frogs and switches.
Journal-bearings which are worn so as to produce excessive side-motion should also be reported whenever they can be detected.

There is a well recognized tendency on the **Cause of**
part of trainmen to account for any delay or **damage.**
damage to a train while it is in their charge by
assigning it to a cause beyond their control;
occasionally from a disinclination to take the
trouble to find out the real cause and at other
times to escape blame. The situation of the
track force makes them particularly liable to
these charges, and for this reason if for no
other the roadmaster should require, and the
section-foreman should furnish, a short but
exact account of any unusual occurrence to a

train which might in any way be charged to track.

Any track which is not considered safe for trains running at full speed should be protected by caution signs or by a slow order posted on the bulletin board at division headquarters; but since engine men are known at times not to regard these means as carefully as they should, caution signs and slow orders should be resorted to as little as possible. When they are used any neglect of them should be at once reported.

CHAPTER III.

FENCES, HIGHWAY CROSSINGS AND PLATFORMS.

A well-fenced right-of-way, although an expensive thing to construct, is most assuredly a desirable thing for a railroad, since the amount paid for damaged stock is usually large and does not seem to be a judicious outlay on general principles. Good fences are easily built and will last seven or eight years without much attention beyond the occasional straightening of a post, nailing on of a board, or tightening of a wire. *Fences desirable.*

The greatest enemy of wooden fences is fire, and on this account the ground around and under them should be kept free from undergrowth and long grass. A simple means of doing this is to plow a furrow close to the fence and on each side of it with the sod turned away from the fence. *Fire.*

The best timber for posts is cedar, since it decays slowly and holds the nails and staples well. Chestnut and white oak are also good, but decay more rapidly. In fact almost any wood which will make good crossties is suitable for fence posts but whatever the timber used, it should always be stripped of its bark before being planted. For the stretchers, pine or hemlock boards, 1 in. x 6 in., were until a few years ago almost universal. At the present time, however, steel wire has taken the place of lumber. For ordinary right-of-way fence, the most suitable form is "woven" or *Posts and stretchers.*

20 · THE NEW ROADMASTER'S ASSISTANT.

Wire fences.

laced in rectangular or triangular shapes, or else stretchers of twisted ribbon or rope, of which there are numerous kinds on the market, varying little in price. Several forms are illustrated in figs. 1, 2 and 3, all of which are

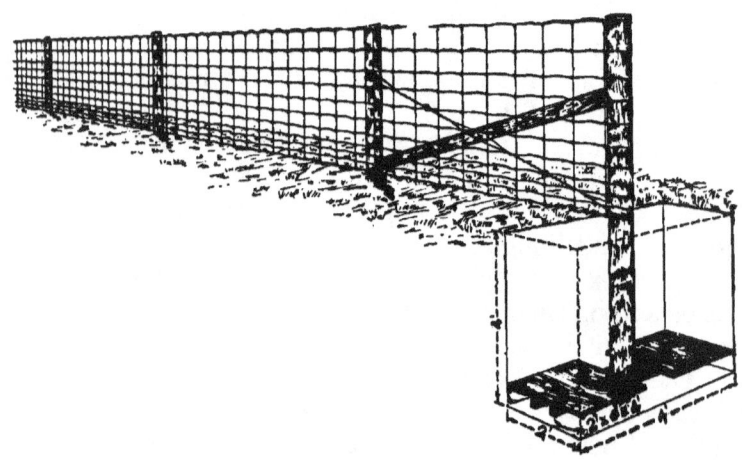

FIG. 1.—Page Woven-Wire Fence.

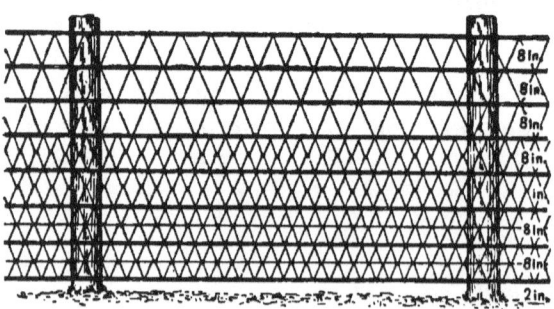

FIG. 2.—Ellwood Woven-Wire Fence.

FENCES, HIGHWAY CROSSINGS AND PLATFORMS. 21

Fig. 3.—McMullen Woven-Wire Fence.

of the kind known as "woven"; a kind which seems to be rapidly increasing in use since it is said to be able to turn all kinds of stock, from the largest to the smallest, without injury to them or to the fence.

The old-fashioned barbed wire is illustrated in fig. 4, together with a simple and useful **Farm gate.**

Fig. 4.—Barbed Wire with Pine Stretchers and Farm Gate.

style of farm gate which needs for its construction nothing but a saw, hammer, nails and boards, while a chain and padlock make the very best fastening possible.

For station grounds and for the right-of-way through towns, the product called "expanded metal," see fig. 5, formed of a steel plate that has been punctured and spread apart, makes an excellent and permanent fence.

FIG. 5.—Expanded-Metal Fence.

Metal posts. Steel or iron posts are used where durable wooden posts are expensive, and some forms of them make a handsome and substantial fence. Figs. 6, 7 and 8 show convenient

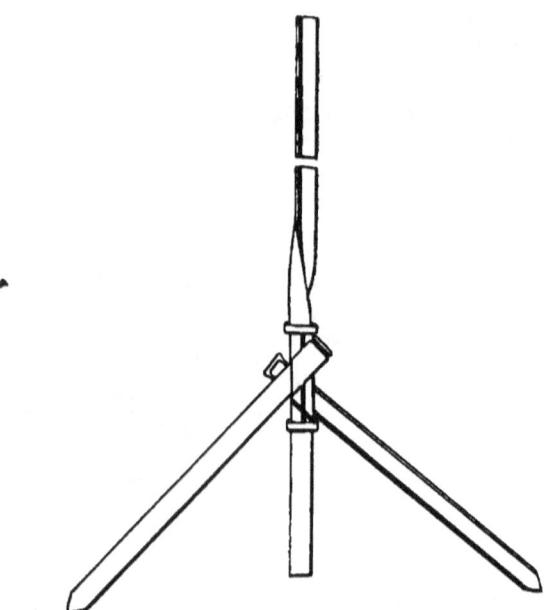

FIG. 6.—The Anchor Post.

and neat forms of posts. Fig. 6 is made of 1¼ in. T-iron, twisted at the ground line and provided with blades which are driven into the ground and serve to brace the post in a

Fences, Highway Crossings and Platforms. 23

direction at right angles to the line of the Metal fence. Figs. 7 and 8 are formed from a thin posts.

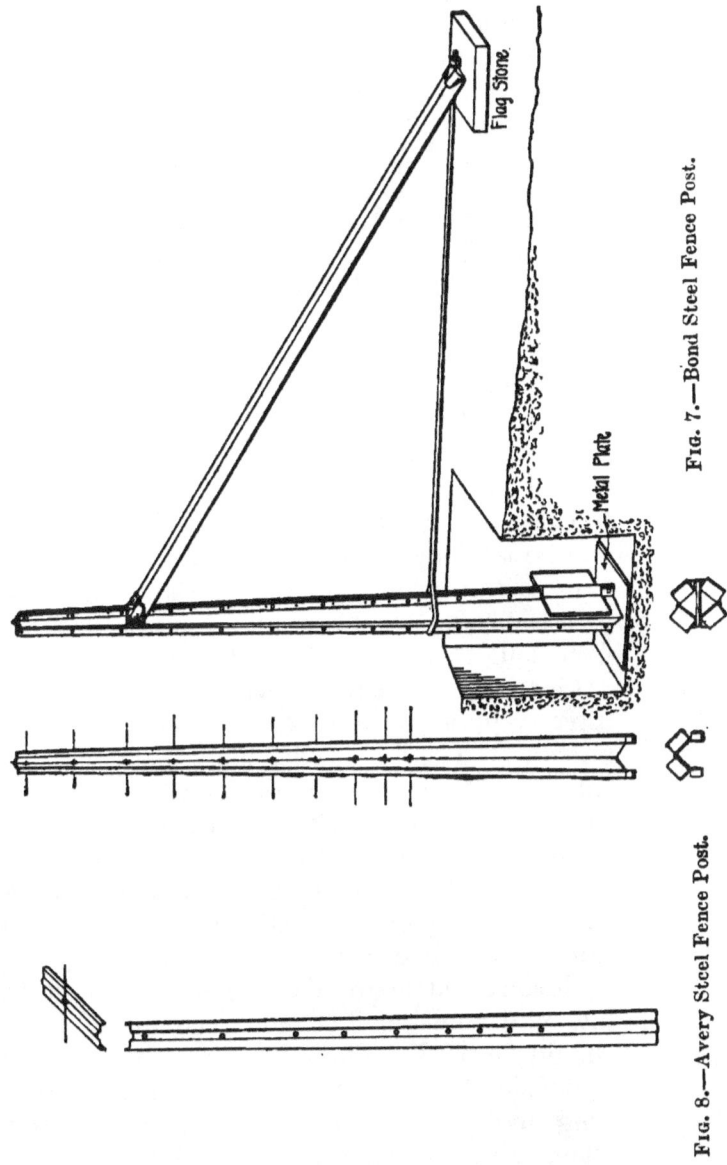

Fig. 7.—Bond Steel Fence Post.

Fig. 8.—Avery Steel Fence Post.

sheet of steel. Of the two posts shown in fig. 7, the one on the left is the ordinary form, while the other one is used at corners or gates and is substantially braced.

Post distances. When boards are used for stretchers the posts should be placed 8 feet apart, but with wire fence this space may be safely increased to 12 feet; the latter distance, however, should not be exceeded, and in the best barbed wire fence, a board, (see fig. 4) 1 in. x 6 in. is used as the top line and nailed on the side of the post for the purpose of attracting the attention of cattle, who might otherwise fail to notice the wire. The end posts of each break in a wire fence, whether road-crossing or gate, should be braced as shown in figs. 1, 4 or 8, to enable them to withstand the pull of the wire.

Fence gang. Ordinary repairs to fences may be made by the regular gang of section men, but for extended improvements, additions, or general repairs, it will be found that much cheaper and better fences can be built if a special fence gang is employed. This gang should have its own tools and hand-car and should consist of from four to six men and a foreman.

Cattle guards. Open cattle-guards, with the rails laid directly upon the stringers, are no longer admissible. They are extremely dangerous to trains, as in case of a derailment they will surely cause a wreck. Neither are cattle-guards desirable with the ties laid upon stringers and an open space underneath, for cattle are frequently caught in them and are killed, sometimes wrecking a train.

Surface cattle-guards are now allowed by law and some one of the recognized forms should be used. The particular kind decided upon should not be easily damaged by dragging brake-beams, etc., should have interchangeable parts and should be susceptible of easy and quick repairs. Several different forms are illustrated in figs. 9 to 14 inclusive.

FENCES, HIGHWAY CROSSINGS AND PLATFORMS. 25

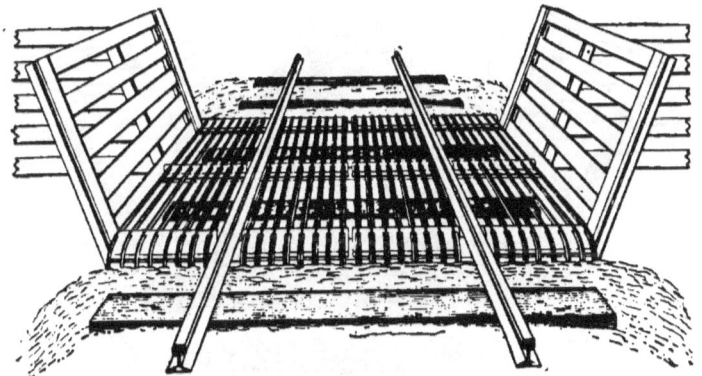

Fig. 9.—Bush Cattle-Guard.

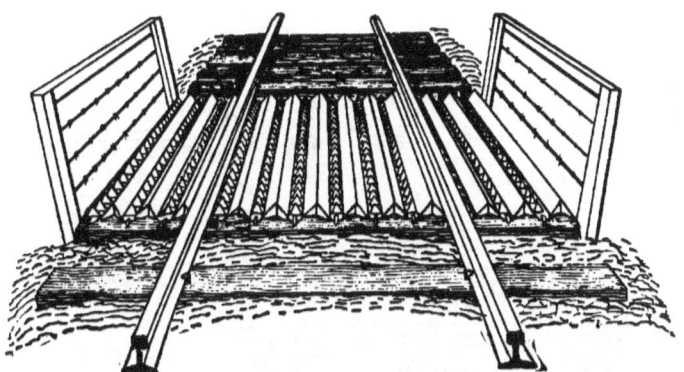

Fig. 10.—Kalamazoo Cattle-Guard.

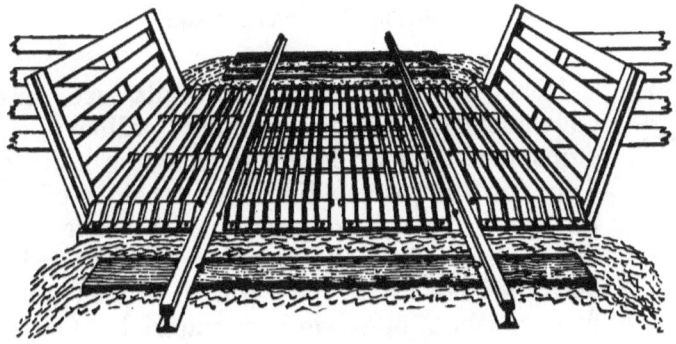

Fig. 11.—National Cattle-Guard.

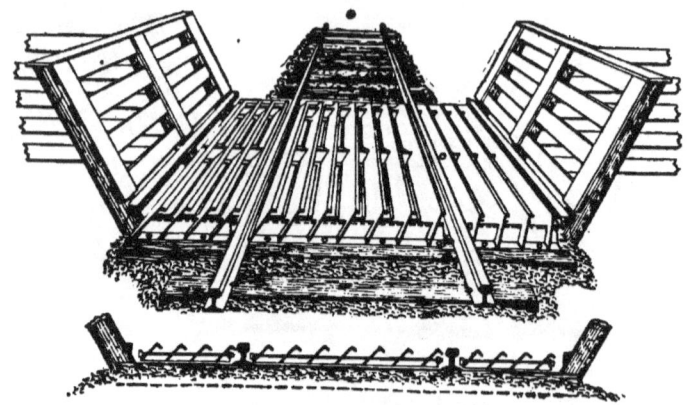

FIG. 12.—Standard Cattle-Guard.

FIG. 13.—Merrill-Stevens Cattle-Guard.

FIG. 14.—Merrill-Stevens Cattle-Guard.

Highway crossing construction. Road-crossings, as usually built, are most conveniently maintained under ordinary circumstances when formed of a frame built of 4 in. (or 5 in., depending on the height of the rail) x 10 in., yellow pine, (never white oak, which warps badly), chamfered on the inside and filled level with the top with broken stone, furnace slag, or a similar material. On the outside of the track, planks should be laid parallel with and close to the rails, and sloped away from the track to meet the paving or dirt of the road. Fig. 15 illustrates this idea.

Highway crossings.

Fig. 15.—Open Highway Crossing.

In cities, or at any highway crossing where the teaming is heavy, a piece of old rail placed between the main rail and the timber, in such manner as is illustrated in fig. 16, will add

Fig. 16.—Old-Rail Protection for Highway Crossings.

greatly to the life of the crossing-plank by protecting it from the grinding of ice and dirt during the passage of trains, while it also facilitates the cleaning of the flangeway.

In laying timber at highway crossings, platforms or at any other place where it is exposed to the wet, it should always be placed as in fig. 17A, in which it is seen that the dip of the grain tends to shed the water and not as in 17B, where it is evident that water would be absorbed and held for a considerable time.

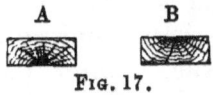

Fig. 17.

The drainage of highway crossings is of especial importance and should be accomplished by running small tile or broken stone drains under the ends of the ties, communicating

Highway crossing maintenance.

with the nearest waterway. Great care must be exercised in maintaining highway crossings, because an injury caused to a horse or vehicle is soon heard from in the shape of a demand for damages.

Terra-cotta flooring. Within a few years, what is to modern engineers at least, a new material has been developed and apparently perfected. It is composed of clay, baked very hard and is used in the form of tiles, as a substitute for asphaltum, cement and flag-stones. For station platforms it is the ideal material, since it costs little if any more per square foot than oak planking, is nearly indestructible and does not decay. If carefully laid in the proper sizes at highway crossings it would furnish a permanent and even surface at an ultimate cost much less than that of timber and with great satisfaction to the public. Since it can easily be taken up and relaid by trackmen of average intelligence, the repairs to track would not be interfered with or made more expensive than they are now at such places.

Station platforms. Platforms at unimportant stations are easily constructed and prove quite satisfactory when built of coarse broken stone for a foundation, dressed and surfaced with crusher dust, clean gravel or cinders from locomotives. The last is somewhat dirty but otherwise is very good for the purpose. The foundation should be well drained and if the platform is maintained with reasonable intelligence it will last a long time at a nominal cost for repairs.

Roads at stations. Driving roads beside unloading tracks may be easily constructed after the old-fashioned "corduroy" plan, by the use of cross ties, the end sills and bolsters of cars and other material of a like character, which through partial decay or for some other reason is no longer suited for its original purpose. A cobble stone pavement laid on six inches of sand, properly drained, is still better and is one of the best of the cheap pavements so far as cost and permanency are concerned. In fact, the usual conditions surrounding places of this kind are so bad as to suggest the idea that almost anything which could be done would improve matters, and it is therefore recommended to trackmen that there is an opportunity in many localities for effecting a change which will prove a great benefit to the patrons of the railroad and indirectly to the railroad itself. *Roads at stations.*

Like the roads beside unloading tracks, the approaches to most stations, except those in cities and their immediate suburbs, are usually in a disgraceful condition. For this the railroad companies can hardly be blamed since the country roads are little, if any, better. But a new sentiment is being aroused, and the farmers are learning that cheap highways are more costly in the end than good ones. Laws favoring the construction of good roads are being made, and the maintenance-of-way force on railroads must be prepared to meet the demand for better roads in station grounds. *Macadam roads are in every way the best for the purpose until the travel becomes great enough to demand a regularly paved way. The drainage question is here, as in so many other places, of the utmost importance. Sufficiently drained, a macadam road may be easily maintained under a heavy travel; insufficiently drained, it cannot be kept up without considerable expense under even slight use, for the frost alone will destroy it in time. *Good roads.*

---

* For a most instructive short treatise on this subject, read "Roads and Pavements in France," by Alfred Perkins Rockwell. Published by John Wiley & Sons, New York.

CHAPTER IV.

MISCELLANEOUS FIXTURES AND STATION GROUNDS.

In the location of all buildings, signals, high platforms, etc., nothing should be placed nearer the main track than 5 ft. from the nearest rail. Although a less distance will clear upon straight track, the 5 ft. leaves little to spare at the top of a car when the outer rail is elevated six inches. *Clearance.*

The section-house should be located in such a way that the car may be got out at any time without the likelihood of being hemmed in by standing cars. It should have work-benches at the sides, with racks at the end and in the ceiling for storing tools. A vise, a draw-knife for shaping handles, a carpenter's cross-cut saw, a two-man saw for platform and highway crossing planks, one or two coarse files, a triangular file, a brace and bits and a grindstone are the tools which, besides the ordinary track tools, are essential in every section-house, although others will be needed in special cases. *Section-house.*

The house should be not less than 12 ft. square and may well be 16 ft. square since it costs but little more and is then quite large enough for several men to work in while the hand-car and push-car are housed. The appearance of the house (fig. 18) should be neat but quite plain, while there should be windows at the sides and at the end opposite the door. The door should be made in two pieces and slide into the front wall.

Fig. 18.—Section-House.

Bumping posts. Although bumping posts are not strictly a part of the track, roadbed or buildings, trackmen will often be required to furnish some means of stopping cars at the end of a track. For ordinary purposes, where nothing but the wild car itself will be damaged, if it runs beyond the end of the track a few feet, the methods exhibited in figs. 19 and 20 will do

Fig. 19.—Curved Rail Bumper.

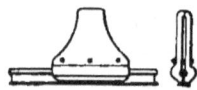

Fig. 20.—Clamped Bumper.

quite well. Those arrangements shown by figs. 21, 22 and 23 are effective when they are

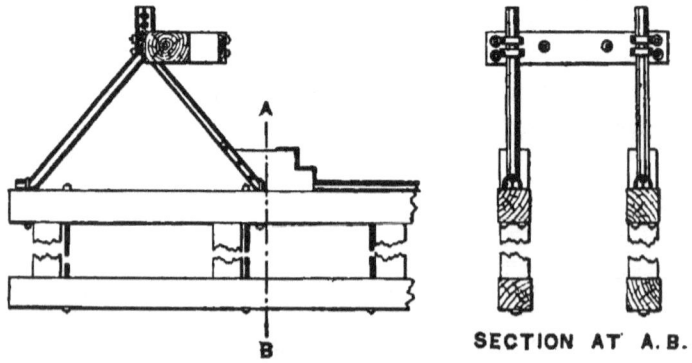

Fig. 21.—Triangular Bumper.

FIXTURES AND STATION GROUNDS. 33

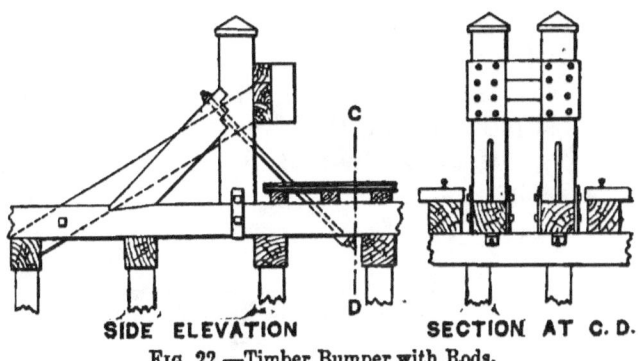

FIG. 22.—Timber Bumper with Rods.

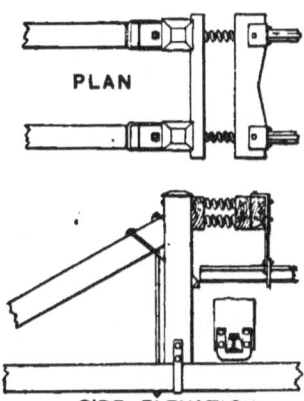

Bumping posts.

FIG. 23.—Braced Spring Bumper.

made of heavy timbers with sufficiently strong tie-rods. The device (fig. 24) rests upon ma-

FIG. 24.—Ellis Bumping Post.

sonry and is expected to stop nearly anything which is likely to be run against it, since the blow is changed from a horizontal to a vertical motion, and the force is largely absorbed by the earth instead of by the apparatus itself.

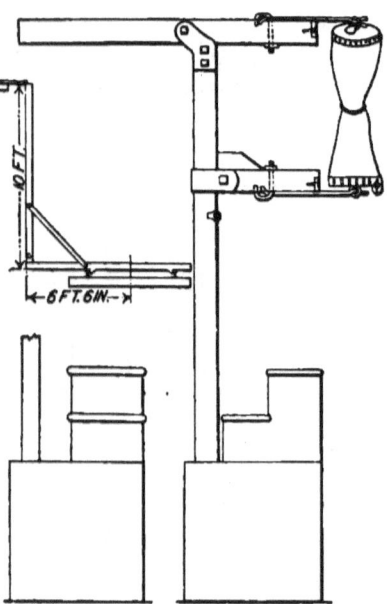

FIG. 25.—Ordinary Mail Crane.

Mail cranes. The mail bag deliverer (fig. 25) is of the simplest, both in idea and construction, and is typical of the usual method. In the upper left hand corner is shown a gage for placing the arrangement, giving the distance of the upper hook above the rail (10 ft.) and its distance from the center of the track. In fig. 26 there is illustrated an invention which is for the purpose, both of delivering a mail bag to a train and receiving one from a train at the same time. It is but a slight amplification of fig. 25, and is of great use in avoiding the danger which must always accompany the *Mail cranes.*

FIG. 26.—Crane for Delivering and Receiving Mail.

throwing of mail bags from a rapidly moving train. The Post Office Department furnishes plans of acceptable mail-cranes, and this is probably the best source of information.

Highway crossing signs, whistle signs, mile posts and other notices put up to attract the attention of the public or the servants of the railroad company, should be conventional in form, that is, of a commonly known pattern, and all signs for a certain purpose should be alike in size, shape, color and in position with regard to the railroad track. *Track signs.*

The sign indicating an approach to a grade crossing with another railroad is only needed when there is no interlocking plant. Its best form is like fig. 27, which has arms 4 ft. long by 6 in. wide with 4-in. letters. The post should be 10 ft. long above the ground and 6 or 8 inches square. *Grade crossing sign.*

THE NEW ROADMASTER'S ASSISTANT.

FIG. 27.—Railroad Crossing Sign.

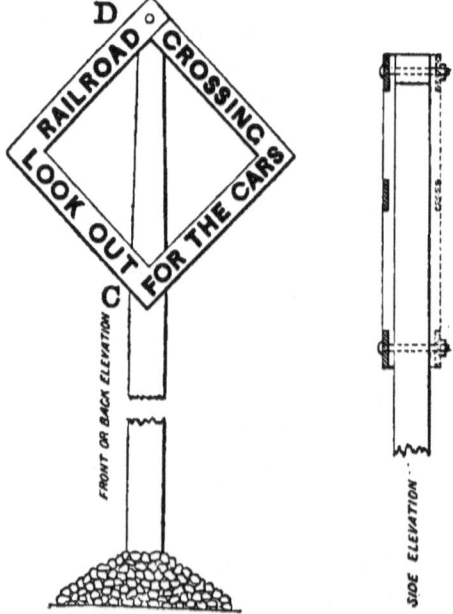

FIG. 28.—Highway Crossing Sign.

The "highway crossing" sign (fig. 28) is a **Highway crossing sign.** conventional one, which has boards about 5 ft. long by 6 in. wide with 4-in. letters. The post should be not less than 12 ft. long above ground with an end section of about 8 in. by 8 in. but tapered in the manner shown between the points C and D. Occasionally it will be found convenient to provide both sides of the post with a sign, in which case the appearance would be as in the side elevation, where the dotted lines represent the second sign. In many states, however, the form and lettering of these signs is prescribed by law.

Fig. 29.—Various Signs.

In fig. 29 are illustrated various signs of the **Various signs.** same pattern but different letters, which may be either raised or not. The base is an iron plate, $\frac{1}{4}$ in. thick, with a border, $\frac{1}{4}$ in. thick and $\frac{1}{2}$ in. wide, and it is mounted on a post, 6 in. by 6 in., 5 ft. above ground. Of those which do not explain themselves in fig. 29, "W X" means "whistle for road crossing;" "W S" means "whistle for station."

The "mile post" (fig. 30) is the well-known **Mile posts.** form, made of a piece of 8-in. by 8-in. timber, 8 ft. long, 3 ft. of which should be in the ground. It should be placed with an edge toward the track, by which means the distance to each terminal may be seen at the same time.

Mile posts. Wherever possible all of the posts should be on the same side of the track.

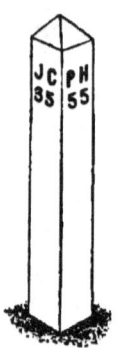

FIG. 30.—Ordinary Mile Post.

FIG. 31.—Mile Post.
(Bond Steel Fence Post Company.)

Metal signs. The use of metal for signs increases their cost, but they defy decay and the guns of the daring huntsmen who use the track as a highway and the sign as a target.

Fig. 31 exhibits a mile post built entirely of metal, a material which is to be recommended for all purposes of this sort where permanency is of value. It can be used to advantage on the signs shown in fig. 29 as well as for the mile post or the monument (fig. 32), which last *Monuments.*

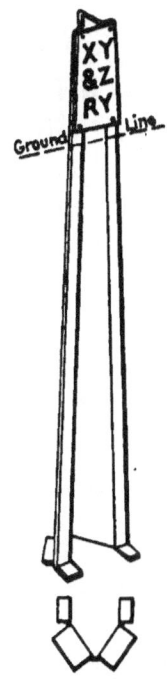

Fig. 32.—Monument.
(Bond Steel Fence Post Company.)

is intended for marking boundary lines or the intersection of the railroad with town, county, or state lines. It is well to mention here that these intersections should always be permanently marked and also the corners made by changes in the width of the right-of-way.

On some railroads these sign-posts are finished by a heap of round stones, a little smaller in size than a man's fist. Sometimes a heap is conical, as in fig. 28, and sometimes it is hemispherical in form. This heap is usually whitewashed, and is for the purpose of adding to the appearance of the post and to prevent the growth of vegetation, which exposes the post to damage by fire. It is a somewhat expensive plan and the benefits hardly justify the expense.

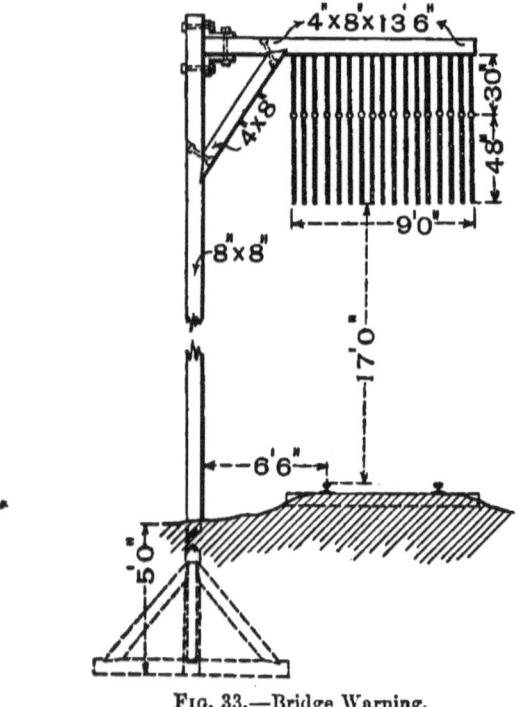

FIG. 33.—Bridge Warning.

Bridge warning. Fig. 33 is a "bridge-warning," and is nearly self-explanatory. The hanging ropes are each $\frac{3}{8}$ in. in diameter, served with twine, not knotted at their lower ends and attached to the crane by $\frac{1}{4}$ in. round rods.

Sign letters. The most permanent letters and figures for wooden boards or posts are made of very thin cast-iron and they are probably cheaper in the

FIXTURES AND STATION GROUNDS. 41

end than letters or figures which must be renewed by a regular painter. Next to them in permanency come those of a good quality of black graphite paint, applied with a brush. This paint should always be used for the black on woodwork.

The fences at road-crossings should be whitewashed at least once a year, since a whitewashed fence may be seen at a considerable distance at night and it is important that enginemen shall know when they reach a highway. Salt used in whitewash causes it to flake off rapidly and should therefore never be used under any circumstances.

Whitewash.

The white paint used on switch targets, signal blades, etc., should be the best white lead. The red paint intended for the same purposes should be real English vermillion, while both white and red should be mixed in linseed oil. Any odor of kerosene or gasoline in an oil paint, means that it has been adulterated.

Paint.

For the iron work of water cranes, switch stands and sign posts, an honest black asphaltum varnish is the best.

Two methods are known, otherwise than by watchmen, for the protection of the public at grade highway crossings. The first is by means of an electric bell (fig. 34), located at the crossing and rung automatically by the approaching train. Usually these bells are actuated by what is called a "track instrument" (fig. 35), which consists of a lever supported by the cross ties, the short end of which is in contact with the rail, as in fig. 35. In other forms it is so placed as to be depressed by the wheels of a passing train. The other end of the lever, by this up and down movement, is caused to make and break the electric circuit in which the bell is placed, every time a wheel passes over the track instrument.

Crossing bells.

Crossing bells. In another arrangement the bell is controlled by what is called the "track circuit,"

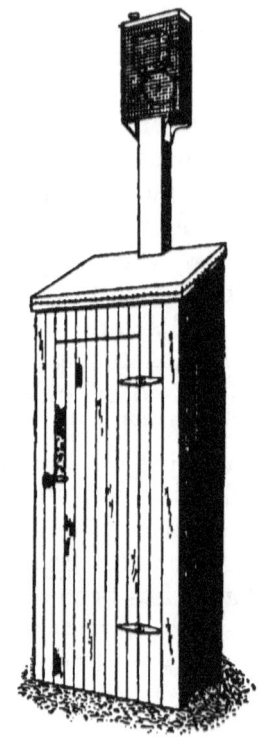

FIG. 34.—O'Neil Crossing Bell.

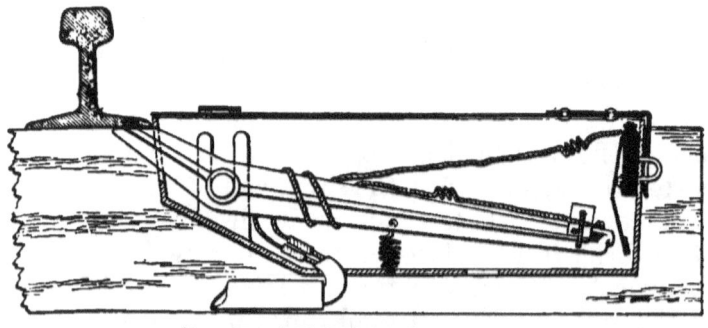

FIG. 35.—O'Neil Track Instrument.

wherein a portion of the two rails of the track form a part of the bell circuit, and the bell is rung by the mere presence of a pair of wheels on the "track circuit." The crossing bell is generally regarded as suitable only at rural or suburban crossings.

The highway crossing "gate" (fig. 36) is the most recent development of this idea which is the one usually applied to busy grade crossings. Originally the gates were operated by means of chains or wires worked from a crank located on one of the posts; but in fig. 36, compressed air is the motive power. For this reason one man is often able to handle the gates at several adjacent crossings, since the pump and valves are usually placed in an elevated cabin from which a clear view may be obtained.

Crossing gate.

In fig. 36, P is the air pump, T - T' the valves which, by their position, determine the gates to be moved and whether they shall be moved up or down. A - A' are the air-pipes, D - D' are flexible diaphragms contained in the ()-shaped air chambers; these diaphragms rest against the plungers R - R' which connect with the cranks K - K' and in turn transmit the motion of R - R' to the sprockets S - S' and the chain C - C'. The gates are directly operated by the system of small cranks and levers which lie immediately above S - S'. The weights W - W are for the purpose of counterbalancing the gates G - G. In the illustration the gates are down, and if it is desired to raise them, T - T' are put in that position which will cause the compressed air to enter A and will open A' to the outside air. The pump is then worked. D (of the post on the right) is pressed toward R, moving R and at the same time K, S, G and C. But C and C' are continuous, and any movement of this chain tends to simultaneously either raise or lower both gates, depending only in which

The New Roadmaster's Assistant.

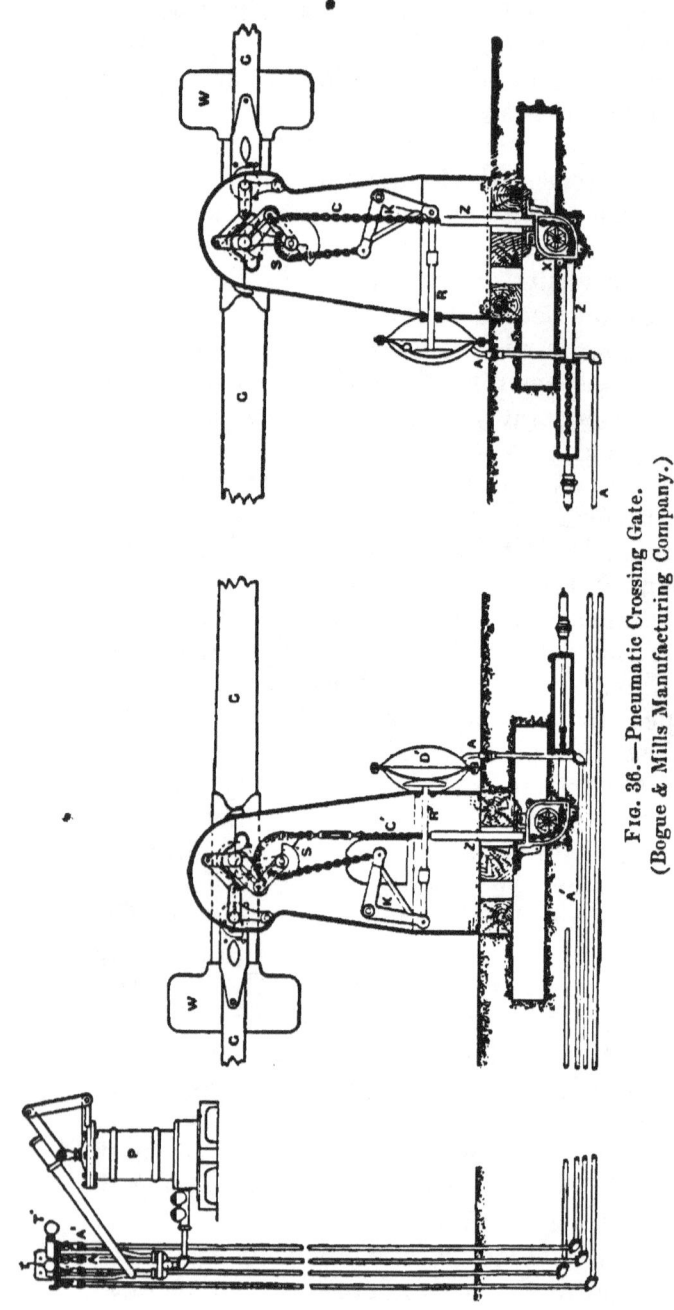

Fig. 36.—Pneumatic Crossing Gate.
(Bogue & Mills Manufacturing Company.)

direction it moves. In other words, when the gates on the right are to be raised, the diaphragm on the right acts for both gates, and when they are to be lowered the reverse action takes place.

Where a street is narrow, one gate on each side of the track is enough, but frequently two on each side of the track are required while a means of still further extending is found by placing the posts on the curb line and using small side-walk arms in addition. These arms are shown broken in fig. 36.

But of all the means for protecting the public at highway crossings the separation of the grades surpasses all others in safety and ultimate cheapness, except in unusually difficult localities. This proposition cannot be too strongly stated since any railroad which, from indifference or any other cause, neglects to efface every grade crossing which they have the power to avoid, is surely nursing some future trouble. *Separation of grades.*

It is not often that too much attention is given to neatness and an attractive appearance on our railroads; on the contrary it is a matter which seems to be regarded, except in a few cases, as of little or no importance. It is not intended to inquire into the reasons for this, but it is necessary to call attention to the fact and to urge an improvement so far as it lies in the power of each roadmaster, supervisor and section-foreman. *Station grounds.*

When the ordinary appearance of a country station is recalled, with its muddy roads, dilapidated fences, dirty platforms, scrap of all kinds kinds lying everywhere in sight, the things which must be done to make the place attractive instead of repugnant and their small cost become evident. The work of one man for a week in a year will usually maintain the fences;

some engine cinders or field stones will fix the roads; a few trees at the borders of the company's land; a little grass plot near the station and some vines at the corners of the buildings will cause the place to look like a gem instead of an open sore. The trees and vines will cost only the labor of transplanting them (elms, maples or oaks, but never fruit or nut-bearing trees, because boys will surely injure them) from the nearest woods, while suitable turf can be secured along the right-of-way or from almost any pasture.

CHAPTER V.

Water Supply.*

The question of water supply is one which does not naturally have a bearing upon maintenance-of-way work, but it is a question which will often be forced upon the roadmaster by circumstances, and a few suggestions here may easily prove of value.

<small>Selection of source.</small>

Given a sufficient quantity at each of several available sources, the only question of importance is as to the quality of the water. It must not (if it can possibly be avoided) carry much free lime, and it should not be muddy. The first condition is most apt to be found in springs and the last in streams, but the lime is a practically incurable fault, while the mud may be much reduced in quantity by allowing the water to settle before finally delivering it to the locomotives. There are many other impurities which render water undesirable, and for that reason it is best to have water subjected to a chemical test before finally arranging to use it, but the two objections already noted, lime and mud, are the most common.

<small>Testing water.</small>

To detect an excess of lime in water it is only necessary to dissolve a piece of white soap the size of a pea in a tablespoonful of freshly fallen rain water. When this preparation is put in a glass of the water to be tested, it will

---

* Many of these notes were suggested by "The Elements of Railroading," by Charles Paine. Published by the Railroad Gazette.

cloud immediately if there is much lime in the water. By using the same quantity of soap-water in several glasses, each holding the same amount of water to be tested, a comparison of different sources may easily be made, for if they contain different quantities of lime those which contain the most lime will appear the most clouded.

Cost of plant. If the samples of water are equally free from impurities then there remains the question of cost. If the water comes from a point 30 ft. or more above the track and is not more than half a mile away (and sometimes even further) then a gravity supply will almost always be found on investigation to be the cheapest and easiest to maintain.

Windmills. Where locomotives take water only at long intervals, a wind-mill may sometimes be used economically and satisfactorily, but they are often out of service for two or three days at a time for lack of wind. So unless there is a very large supply reservoir or but three or four engines a week are to be expected, they cannot be relied upon.

Hydraulic ram. What is known as a hydraulic ram will, where economy of water is not an object, automatically raise considerable quantities of water with practically no attention.

Gas and steam pumps. In point of convenience and amount of attention necessary, a gasoline or kerosene pump comes next to a hydraulic ram. These pumps are fired automatically and stopped by the filling of the tank, and are usually less costly in operation than a steam pump where an attendant is needed for at least part of the time. But whatever pump is used, let it be a good one and of comparatively large capacity for the work to be performed.

Storage reservoir. The storage reservoir must be near the track, its bottom at least 25 ft. above the rails and

WATER SUPPLY. 49

the pipe connecting it with the cranes should be not less than 8 in. inside diameter. Occasionally it will be found that the storage tank may consist of a paved earthen reservoir located somewhere near the right-of-way, in which case it should be covered with a conical or pyramidal roof, to protect the water from leaves, sticks, etc.

Usually, however, a wooden tank mounted on posts will be found necessary. This should be (fig. 37) of frost-proof construction and

Frost-proof tank.

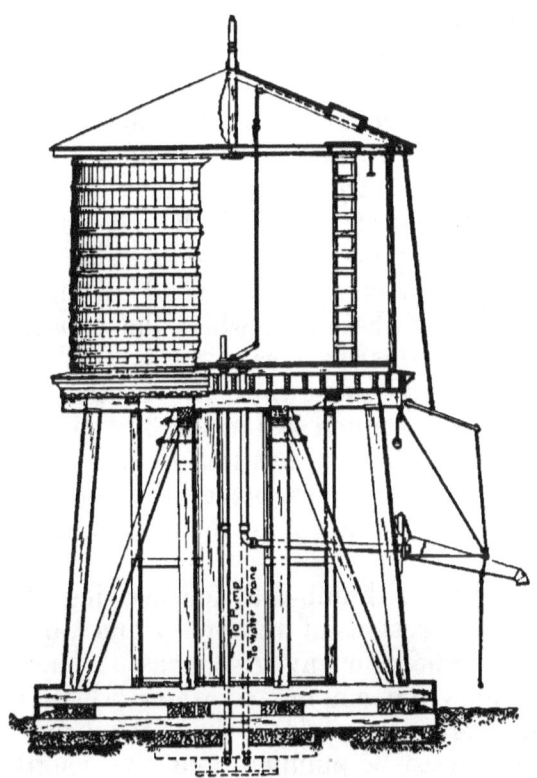

FIG. 37.—Frost-proof Water Tank.

when located at a station should be placed at a distance from the tracks and where it will not have to be moved because of changes. Where the tank stands beside a single-track road a spout may be attached to it, as in fig. 37, but on double track or where the tank is removed from the line a separate water crane must be provided, in which case the spout is omitted.

Reservoir capacity. The storage reservoir, of whatever kind, should contain not less than 25,000 gallons, which would be held by a tub 16 ft. in diameter by 16 ft. (about) high, the common size. This is sufficient to entirely fill the tanks of from six to seven locomotives.

Size of pipe. Pipe which is too small in diameter is frequently used for connecting the source with the storage tank. This is done through ignorance usually, but that does not help to relieve the embarrassment of the situation when it is found, too late, that what should be a full stream of discharge into the tank is nothing but a ridiculous trickle. Although the factory cost of 3 in. pipe is twice as great as that of 2 in. pipe, the cost of fitting and burying them is practically the same, while the capacity of the 3 in. pipe is twice as great as that of the 2 in. At the same time the frictional loss, "choking," is much less in the 3 in. pipe than in the 2 in. This means that on long lines it will take a much more powerful pump to force the water through a 2 in. than through a 3 in. pipe.

Care in designing. There is no thumb-rule for arriving at the correct dimensions of the pump and pipe line which are best for any given case. They depend upon the amount of water which must be delivered in a given time, the height to which it must be pumped, and the length of the pipe line. It may sometimes be cheaper to put in a comparatively large pump and a comparatively small pipe line, but such a case would be very rare, and it is well to stick to the idea of using large pipe; it is also well

WATER SUPPLY. 51

to remember that sharp corners in the line are a considerable obstruction.

In the choice of a water crane there is a considerable opportunity for selection; the points to be considered are size and general arrangement. The size is easily determined; it should deliver a stream of water not less than 8 in. in diameter, which size should continue all the way to the tank. Smaller cranes are built but they deliver water so slowly as to cause annoying delays. The general arrangement of the crane is usually best determined by the reputation of its maker, but one thing

Water cranes.

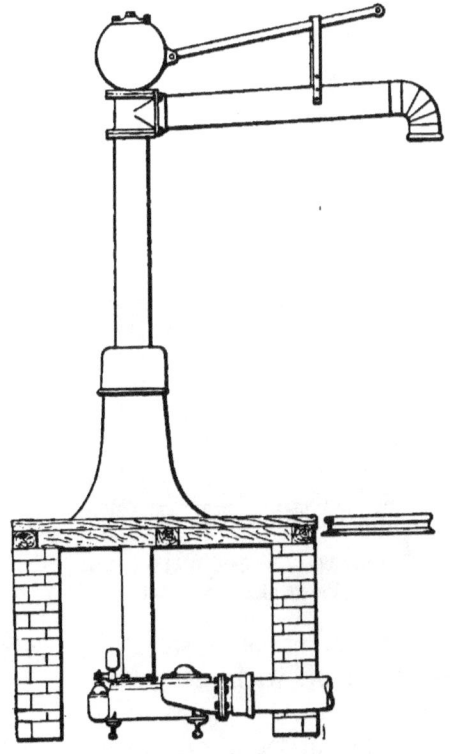

FIG. 38.—Poage Water Crane.

must be borne in mind: the valve must be nicely graduated, for if it is not, when shutting off the water, the pipe line is apt to be burst by a rise in pressure due to a too rapid stopping of the flow.

The ease of control is closely allied to this. Usually it is best to have the valves operated from the end of the crane, as in figs. 38 and 39, which are of standard makes, so that the

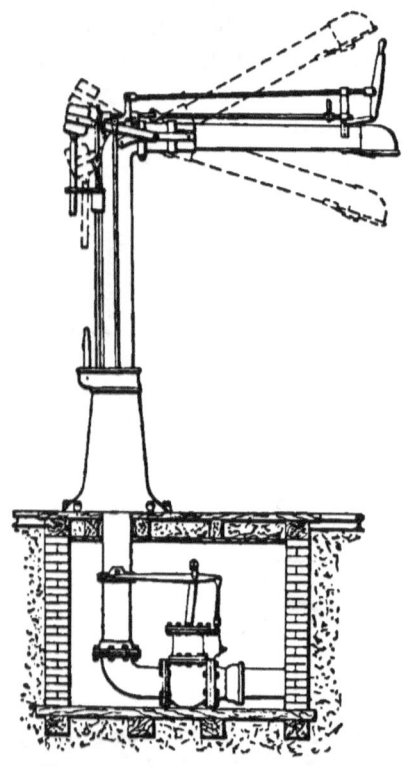

FIG. 39.—Sheffield Water Crane.
(Fairbanks, Morse & Co.)

fireman may start and stop the water and watch his tank fill without getting down from the tender.

Crane pit. The pit should be about six feet deep with stone or brick walls and a cover which contains an air space of four or five inches, the best method of preventing the penetration of frost. There should also be a drain under the

Water Supply. 53

Track tank.

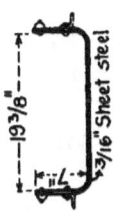

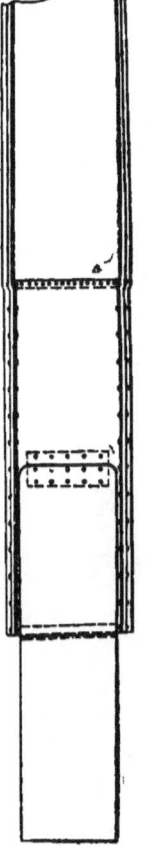

Fig. 40.—Track Tank.

end of the crane to carry off the drippings, and this drain should be easily opened in order to be able to free it of ice in winter.

Track tanks.

Many railroads with fast trains have equipped their high-speed tracks with troughs, by means of which and by scoops that are attached to certain locomotives, the fast trains are not required to stop or even slow down very much when they wish to take water. Fig. 40 is from a drawing of the track tank used by the Michigan Central Railroad, and does not differ much from those used on other lines. The trough is made from a series of plates curved in the form of a flat ∪. It is fed at intervals throughout its length from a frost-proof tank located somewhere near the track, through the 3-in. pipe. This pipe also serves to supply steam to the trough in winter to prevent ice from forming.

CHAPTER VI.

Drainage.

Since the greatest enemy of the track is water, good drainage becomes a matter of the last importance. Without sufficient ditches, the best ballast fails in its office, soon becomes filled with sand or clay, and, in winter when quick drainage is a necessity, acts as a reservoir to hold the water, with heaving track and all its miseries as a consequence. Although in Chapter II it was stated that ditching should be commenced in the spring as soon as the frost is out of the ground, it must be understood that this item of track work is never unseasonable, but should be pushed whenever necessary even to the exclusion of other work, because it quickly and amply repays all of the labor spent upon it. *Time for ditching.*

Each section should be provided with a ditching line, which should always be used in cleaning out an old ditch or opening a new one. Nothing looks worse than a water-way which staggers, now toward, now away, from the track, narrow in some places and wide in others. Both in a flat country and through cuts a cross section of the track and ditch should appear somewhat like fig. 41.

The "berm," or shoulder next to the track, should be lower than the bottom of the ballast at the centre of the track, and sufficiently wide to insure its acting as a support for the ballast, for which purpose it is principally intended. *Section of road-bed.*

THE NEW ROADMASTER'S ASSISTANT.

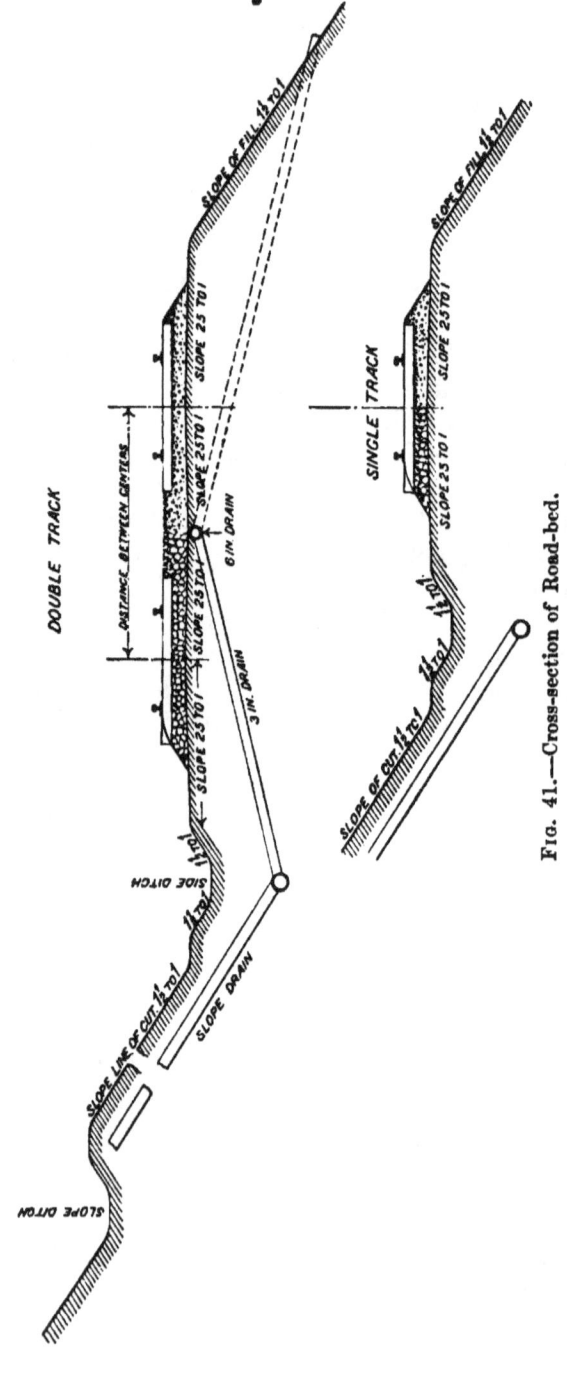

FIG. 41.—Cross-section of Road-bed.

In fig. 41 it will be noticed that all of the corners are rounded, a form which nature will eventually force them to take and consequently the form which should be given to the earth at the start; otherwise the ditches must soon be cleaned of the material which will fall from the edges.

The width of the sub-grade from corner to corner varies greatly on different railroads. On double track, 13 ft. centers, $8\frac{1}{2}$ ft. ties, 12 in. of ballast and a 3 ft. berm—30 ft. would be the width, but if in the practice on any railroad any of these dimensions is different, it is evident that the width of the sub-grade will also be different. On single main track with $8\frac{1}{2}$ ft. ties, 12 in. of ballast and a 3 ft. berm, the width of the sub-grade becomes about 17 ft.

The practice of beginning a ditch at the upper end is so ridiculous that one would suppose it unnecessary to caution trackmen against it. Experience, however, proves the contrary to be the case, since it is a common and serious mistake. *Ditching methods.*

In many heavy cuts, as well as in making new ditches, a ditching plow which can be hauled by the locomotive of the work train will be found of considerable assistance in loosening the earth preparatory to loading it on cars, while, for disposing of more than five or six carloads of waste material, the unloading plow should be brought into play.

Fig. 42 illustrates a complete ditching machine. All the motions of the crane are performed by means of compressed air, while it has connected with it a plow for loosening the earth and a scoop for loading the earth on cars. An unloading plow is also provided, and these three articles are shown at rest upon the crane-car. Where much ditching must be done, *Ditching machine.*

some such device is necessary from the standpoint of economy.

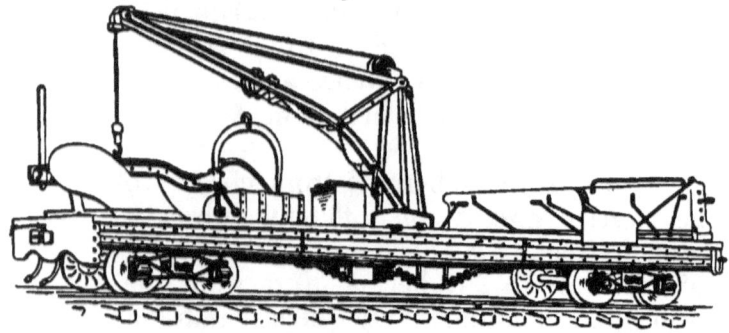

Fig. 42.—Compressed-air Ditching Outfit.
(American Steel Foundry Company.)

Disposition of waste. The material taken from ditches should never be thrown up on the bank where it will be washed down again by the first rain, but should be loaded at once on the work train or the push car and permanently disposed of. In this connection it is well to say that every possible means should be taken to protect the ballast from this waste material which, if it becomes mixed with the ballast, is harmful. For all but the heaviest ditching long-handled shovels should be used, in order that the men may easily reach the top of a loaded flat car from the bottom of a ditch, and also that they shall not be forced to stoop too low in making a thin cut with the shovel.

Large stones are not only unsightly when left in the path of a ditch, but are detrimental. They may be sunk and buried where they lie, blasted and so broken up, or a fire may be built over them until they are very hot, when it is often possible to shiver them by pouring cold water on them.

Paving ditches. In towns, or where the work done is likely to be of a lasting character, it will be found desirable to pave the ditches with large cobble stones, which can often be taken from gravel ballast, where they are always undesirable. Ditches paved in this way, when given sufficient fall, flush themselves during each heavy rain and retain their shape for a long time, particularly when located at the foot of a well-sodded bank or a retaining wall.

Tile drains have been used from the most ancient times, and although their value is well known to many people, they have not until recently been used to a considerable extent on railroads, where, in hundreds of serious cases, they would effect a perfect cure. Tiles should be placed below frost (which varies from nothing in the South to at least five feet in the extreme Northern States), with the ends of the tiles nearly but not quite touching, since they are intended for collecting the water quite as much as for carrying it off. To prevent dirt from being carried into the drain a sod should be turned upside down over each joint and the efficiency of the drain is greatly increased by covering the line of tile with several inches of coarse gravel or locomotive cinders. In wet slopes the tiles should be laid in parallel lines running diagonally down the face of the bank in the direction of the fall of the track, and of a size or frequency depending upon the amount of water to be carried away. *Tile drains.*

The diagonal drains (fig. 41) should be connected at their lower ends to a larger drain laid under the ditch, which should increase in size in the direction of its fall. If a spring exists in the bank a separate line of pipe should be laid from it to one of the diagonal drains, or to the large drain, while to secure perfect drainage on double track another line of tiles must be run between the tracks, just below the ballast, which last line should have frequent outlets communicating with the large drains located on the outside of the tracks.

Varieties of tile. These tiles are made in many forms and sizes and of many different kinds of clay; some are glazed and are therefore quite costly, but between these and the poorest quality are plenty of grades sufficiently good for the work in question and not high in price. A special shovel is made for tile ditching (illustrated in the chapter on Tools) which is the most convenient form for this work.

The desultory way in which a section gang is usually forced to carry on a large piece of work does not lead to economical results; therefore if a large amount of tile is to be laid it will be most cheaply done by organizing a special gang for the purpose, or by letting the job to some outsider at so much per running foot.

Slope ditches. Wherever a cut has higher ground above the slope line, a ditch (fig. 41) should be dug, somewhat above it, to interrupt all surface water which might otherwise flow down and so destroy the slope.

Pole drains. As a substitute for tiles, straight poles roughly trimmed of their branches will serve. They should be laid heads and points, in a bunch of three or four, with their ends slightly overlapping, and at about the same depth as a tile drain. They make an excellent medium and will carry off large quantities of water, but are never so good as tiles.

There are many points around a track, such as highway crossings, wagon tracks at stations, etc., etc., where a ditch cannot be placed, or would be of no use, which might be made perfectly dry by sub-drainage at small cost, instead of being allowed to remain in the bad condition so often seen at such places.

Sodding banks. After a cut has been properly drained its banks should be sown with grass seed, or better still, sodded, as a grassy slope is not only attractive in appearance but it will hold the earth firmly in position. The difficulty in making grass grow at these points is chiefly due to poor soil and a too great steepness of the banks. The only means therefore of obviating the trouble is to supply a good covering of loam on a properly sloped bank. On embankments the turf should be made to grow over and on top of the sub-grade for about a foot, forming a sightly border and affording protection to the shoulder of the bank.

The form of a bank will differ somewhat according to the material of which it is made. The commonest slope is 1½ to 1; that is in fig. 43 (which represents either a cut or fill), the

Slope of embankments.

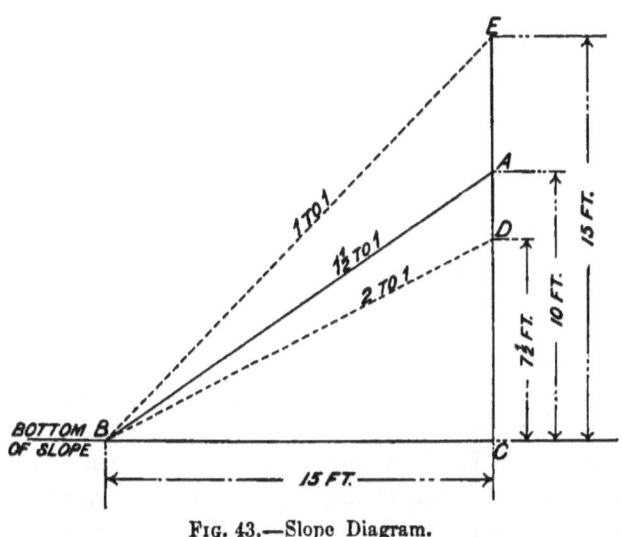

FIG. 43.—Slope Diagram.

middle line is seen to be 10 ft. above C at A and 15 ft. away from C at B. The first distance given, 1½, refers to the horizontal line B C, and the second distance, 1, refers to the vertical measurements C–D, C–A, C–E, which in fig. 43 are 15 to 7½ (2 to 1), 15 to 10 (1½ to 1), and 15 to 15 (1 to 1).

With such poor material as clay or fine sand the inclination may have to be reduced as much as two feet out to each one in height, that is, 2 to 1, while with loose rock 1 to 1 is usually sufficient until, as the character of the ground gradually approaches solid rock, the sides become more and more steep, so that they will finally reach an almost, if not quite, vertical position.

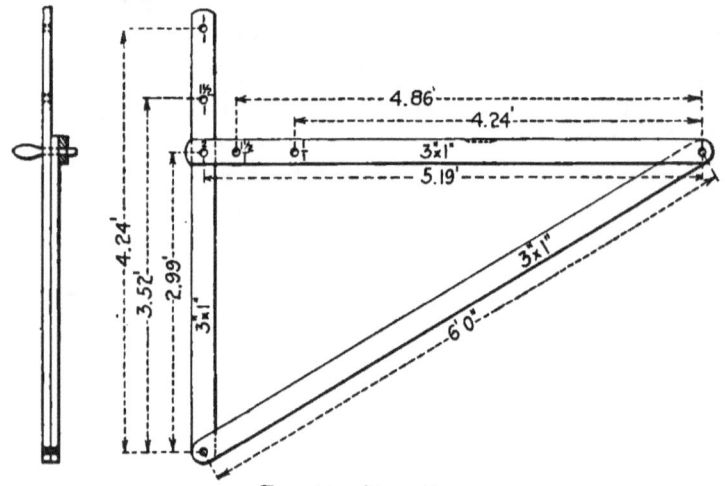

FIG. 44.—Slope Gage.

Slope gage. A gage (fig. 44) for determining the slope of any embankment is a convenient and inexpensive device. It is formed of three pieces of 3-in. by 1-in. pine with a fixed diagonal distance of 6 ft. from pin to pin. The other two pieces have each three holes which are marked $1-1$, $1\frac{1}{2}-1$ and $2-1$; when they are fastened together at these corresponding points by a moveable pin, and the top piece (which is provided with a level) is held level, the diagonal piece will show the slope desired. This tool may be folded up when not in use by simply taking out the pin in the upper left-hand corner. Experience is the only successful teacher in enabling one to previously determine what slope to give a bank, and even the best judgment will occasionally prove at fault.

When an error of this kind is made and a bank is found to be continually sliding there are several ways of treating it. If it is a fill the drainage question seldom enters and the trouble is usually cured gradually, by either sodding the bank, if that seems likely to be sufficient, or by dumping new material as fast as it is required. It will be often found, however, that this cannot be done without buying additional land along the right-of-way, or else paying damages to the neighboring proprietors for the occupation of their lands. When none of these methods will answer, a retaining wall of some kind becomes necessary, which, if it is to be permanent, must be built of stone although it will last for many years if made of old timber or cross ties in the form of a crib. The same conditions hold in cuts except with regard to drainage, which there is apt to be the most important question. One cure has already been suggested in the treatment by tile drains, but cases may be met in which both tile drains and retaining walls will be required. *Preserving embankments.*

If, as occasionally happens, an old embankment begins to slip, it may usually be stopped by placing tile drains in the manner already mentioned or by digging trenches four feet wide and four feet deep every forty or fifty feet, at right angles to the track, and filling these trenches with rubble stone or small nigger-heads. Heaving may sometimes occur on embankments five or six feet high, due to the absorption of water from the bottom, and this also will yield to the usual and useful remedy, the tile drain.

Retaining or "face walls" of considerable size should be built by masons, but smaller *Retaining walls.*

Retaining walls. ones are often needed and these may be readily constructed by the track men, particularly if the stone is conveniently situated and easily worked. Small boulders and loose rock when hammered roughly into shape, carefully laid and well backed, make a good wall. There should be (fig. 45) a firm and reliable bed for

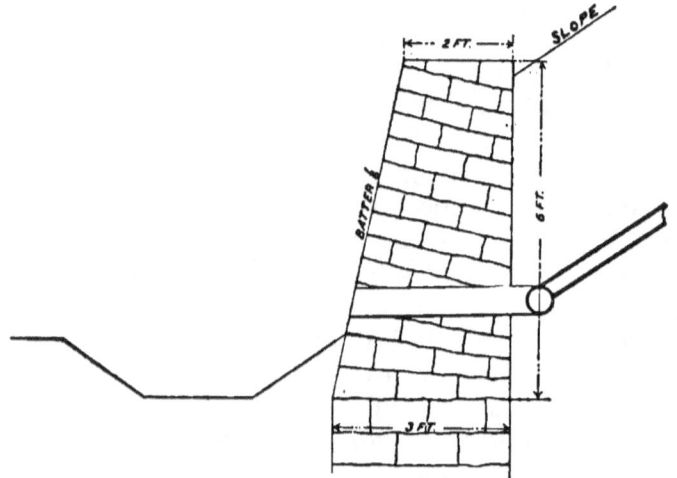

FIG. 45.—Retaining Wall and Ditch.

the foundation, begun below the point to which frost penetrates, and means must be taken to provide a quick and easy passage of the water from the back to the face of the wall. This is best accomplished through loop holes, "weepers," in the masonry every few feet, the bottom of the holes slightly above the high-water line of the ditch, while the rapidity of drainage will be increased by a back filling of coarse gravel or locomotive cinders. At excessively wet places a line of tile drain just back of the wall and connected with the loop holes, will be found of assistance in disposing of the water, and will in that way offer great protection to the foundations as well as to the rest of the wall. The face of the wall should slope from the top to the bottom or have what, in other words, is called a "batter." The amount of the batter will vary somewhat with the circumstances, but for ordinary walls (as in fig. 45) 2 in. for 1 foot in height is sufficient. The base of the wall proper (that is, the top of the foundation) should equal one-half its height, and the top of the wall should equal one-third of its height, so that a wall extending 6 ft. above the ground line would have a base 3 ft. wide and a top 2 ft. wide, as is shown in fig. 45.

The foundation, as will be seen, has no batter and no particular depth is given; this, as has been stated, must depend upon the quality of the bed and the distance to which frost penetrates. The poorest bed for a foundation is composed of quicksand or bog; the best is solid rock or coarse gravel. Between these are different kinds of material, some fairly good, many very bad and since so much depends upon a successful selection it is best to ask advice where the matter seems at all uncertain. Much time and expense may be saved by sinking test holes before the work is begun, which if done will show what may be expected and thus enable one to make all necessary preparations. The resistance of a wall is greatly increased by slightly sloping the stones that lie above the foundation, as shown in fig. 45. *Wall foundations.*

A cheaper, but not nearly so effective, way of holding the toe of a bank is accomplished by laying upon their edges stones which are somewhat flat in form, following always the natural slope of the bank. This method will tend to prevent the surface earth from sliding and may also be used with advantage at the foot of banks that are washed by a stream, particularly where the course of the stream is curved.

CHAPTER VII.

CULVERTS, TRESTLES AND BRIDGE FLOORS.

Use of stringers. The main track should never be laid directly upon stringers. Cross ties should always be interposed to bear the continual pounding of the rails as well as to counteract their tendency to crowd apart. On coal trestles, and other structures of a like character where it is necessary to unload material by dumping it, the rails may be laid upon stringers and these should be frequently tied together by 1-in. round iron rods. Reverse pointed spikes (fig. 46), if any, must be used, for fastening down the rail, as the ordinary form tends to split the stringer. Better than the spike, however, for this purpose, is the interlocking bolt (fig. 47), which

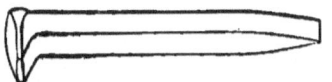

FIG. 46.—Reverse Pointed Spike.

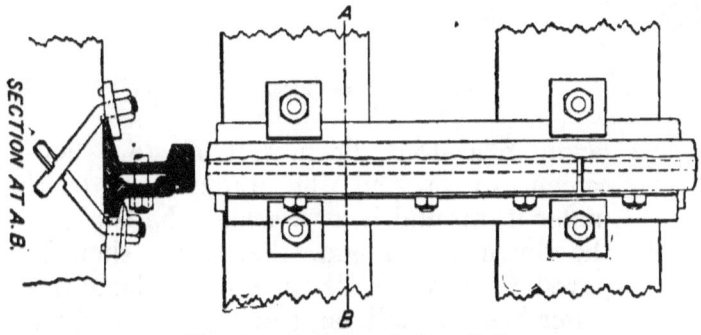

FIG. 47.—Bush Interlocking Bolt.

Culverts.

does not work loose, wear out quickly, or split the stringer, or else a lag screw with a clip to cover the base of the rail (see fig. 112, Chap. X).

The use of wood as a support beneath the ties of a main track cannot be recommended. It rots and it burns, two faults not shared by steel, which is now so cheap and is rolled in so many varying shapes as to adapt it to almost all kinds of work. When therefore, an open culvert is unavoidable it is best built with stone walls, iron stringers and a standard bridge floor. In small openings, I-beams tied together at the ends and well braced may be used where the culvert cannot be covered in the form of a box or an arch.

Cast-iron pipes are the best material for small culverts, and up to an end area of about fifteen square feet are cheaper than an arch. Less substantial, but cheaper than iron, are baked clay pipes, which are now widely used with good results. Both kinds may be laid singly or in numbers, one beside the other. At the upper end, such a culvert (fig. 48), should

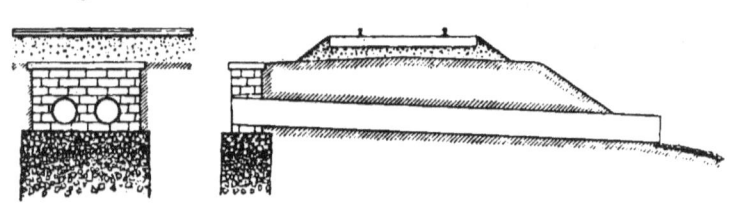

FIG. 48.—Pipe Culvert.

begin in a stout, deep wall of stone laid in cement to prevent the water from leaking under it, around it, or washing the bank. At the down-stream end, the outlet should be paved for a short distance, particularly if the water has any fall on leaving the pipe, while the earth under the whole length of the pipe should be rammed and well settled before the pipe is laid.

Old wooden culverts can frequently be repaired and made permanent by inserting through them cast-iron or earthenware pipes, afterwards filling around them with dirt well rammed in. It is well to remember that the material for this filling should be the same as that which constitutes the rest of the bank up to the bottom of the ballast, in order that the rate of heaving in winter shall remain the same.

For covered waterways a stone arch is by all means the best form of culvert, although a strong, flat, stone cover, where the span is very short, will do quite as well.

The theory that it is best to endeavor to re-rail a derailed truck before it reaches a bridge has given way to the practice of building floors in such a manner as to carry a derailed wheel across the bridge without causing a wreck. This last plan is sometimes accomplished by placing the ties very close together, or by putting heavy iron plates on each side of the rails and on top of the ties or both, thus providing a nearly smooth floor along which a flange may travel without much shock or jar. A modern method, concerning the economy of which opinions differ, is to provide a floor system which will permit the standard track, including ties and ballast, to be carried entirely across the bridge. On metal bridges this is done (usually) by means of what are called "buckle plates," and since it is a method entirely beyond the powers of the track force, it need not be discussed here. *Bridge floors.*

Fig. 49 illustrates one form of ballasted wooden trestle whereon it will be seen, by comparison with fig. 52, that the only essential additions to the ordinary trestle are four stringers, a floor of planking and a curb on each

Bridge floors. side of the planking. The danger of fire is almost banished, the cost of maintenance should be reduced and the life of the trestle prolonged. On the Louisville & Nashville

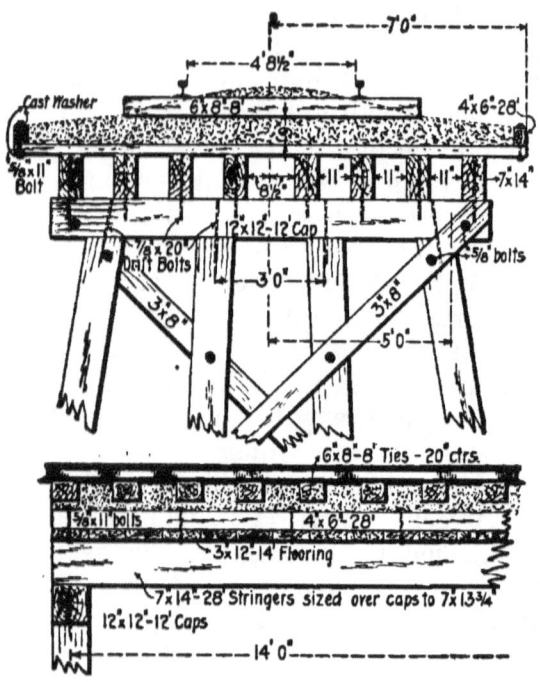

FIG. 49.—Lee's Ballasted Trestle.

Railroad the flooring planks of such trestles are creosoted in order to preserve them.

Re-railing. In fig. 50 it is proposed to re-rail a pair of trucks which have left the rails, first by forcing them into the straight position by means of the outside guard timbers (which are shod with iron plates) and the inside guard rails; then, by means of the castings A and B, to raise the wheels so that those outside will be carried up to the top of the rails and gradually pulled into place. The inside guard rails are often finished off and brought together some distance from the bridge with an iron

CULVERTS, TRESTLES AND BRIDGE FLOORS. 71

point, taken from a condemned frog. It is **Re-railing device.**
doubted whether this is a good plan since, in

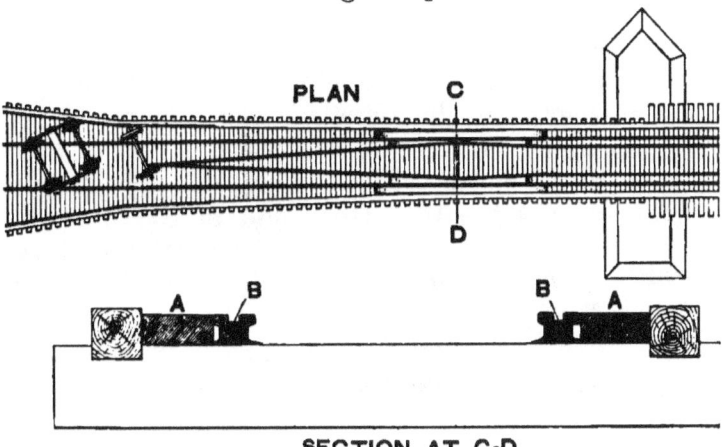

FIG. 50.—Bridge-approach Re-railing Device.

many cases, the wheels of a derailed car are
diverted more than half the gage of the track
and the pointed guard rails then become a
source of danger.

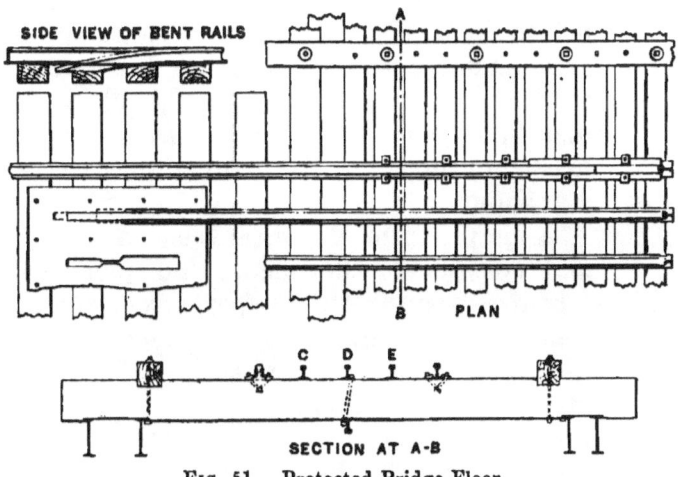

FIG. 51.—Protected Bridge Floor.

Fig. 51 illustrates a method, used on some **Bridge floor.**
roads, of providing a nearly solid floor but

with no attempt to replace any derailed wheels. The central rails C – D – E are bent down at the ends of the bridge so that nothing will catch upon them, and the ties are placed closely together in order that the shocks to the derailed wheels may be very slight. It is evident that a strip of heavy metal plates laid on each side of the main rails will still further add to the protection of the ties and the efficiency of the device.

On all bridge floors, where the ties are laid directly on the stringers, the ties should be frequently bolted to the guard timbers on the outside of the track, and inside guard rails should be provided, securely braced and fastened to the ties.

Shimming bridge floors. Trackmen should be very cautious about shimming the ties or stringers at bridges, trestles and culverts. The practice is much overdone and should by all means be left to the bridge gang, except in cases of extreme necessity.

Trestles. The construction of trestles must usually be left to the bridge gang, but occasions will often arise on small roads where the knowledge as to how a temporary trestle should be built may be of considerable assistance to the roadmaster in repairing the road after a wreck or a washout.

A simple form of trestle is shown in fig. 52, and consists of abutments and piers called "bents," spaced 12 ft. apart from center to center. These support the "stringers" on which the cross ties and rails are laid. The top and bottom pieces of a bent are called respectively the "cap" and "sill," the outside inclined posts are called "batter posts," while the inside posts are called "plumb posts." All of these pieces should be formed of sound timber and, except the stringers, of not less than 12

CULVERTS, TRESTLES AND BRIDGE FLOORS. 73

in. by 12 in. square, nor more than 20 ft. long, **Trestles.** because this method is not adapted to higher

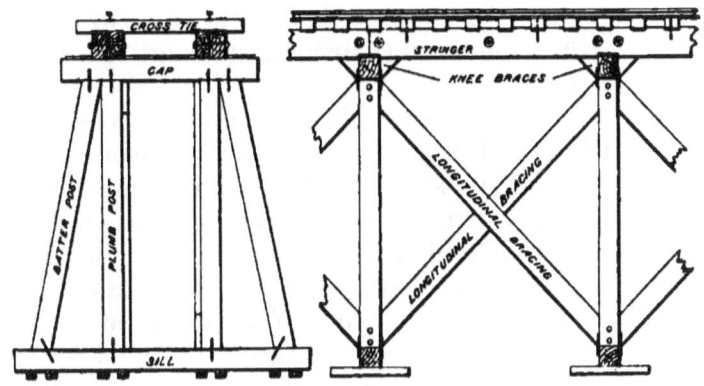

FIG. 52.—Typical Trestle.

structures. The stringers, if formed of clean white pine, should be four in number, two under each rail, and with an end section of 8 in. x 16 in. Either the size or the number of stringers must be increased if the material is not perfectly good. Wherever it is possible the joints of the stringers should be broken, but this will require 24-ft. timbers, which cannot always be secured when they are needed, and the necessity may be avoided in temporary work by substituting strong knee braces overlapping the joints and nailed firmly to the stringers with heavy boat spikes.

If the trestle is likely to remain long in place, the posts should be fastened to the cap and sill by 1 in. drift bolts 2 ft. long. Parallel stringers should be joined by 1 in. bolts and separated from each other for the purpose of drainage by washers 1 in. thick, while to prevent them from shifting sideways a two-inch plank may be nailed along the top of the cap, close up to and in contact with the stringers. The stringers must also be securely braced against the bank at each end in order that the trestle shall not lean. Where there are more than three or four bents, longitudinal braces of 3 in. x 12 in. material must be provided to stiffen the trestle through its length as shown in fig. 52.

Erecting bents. In rapid streams it is sometimes necessary to float the bents into place, but this can usually be accomplished by means of an anchor line up stream and eight guy lines, two at each end of the cap and two at each end of the sill. Care must be taken that the sill shall rest on an even foundation and, if possible, one whose material, when exposed to the wash of a stream, will not scour from under the sill and let it sink. To prevent this it is well to dump some large stones and brush at the sides of and at the up-stream end of each bent, after it is in place, particularly if the trestle is to remain for any length of time.

CHAPTER VIII.

Ballast.

The usual reasons given for not having good ballast under a track are that it cannot be found near enough the place where it is wanted, or that the road is too poor to get it when it is close by. Ninety-nine times out of a hundred the first reason is a wrong one. Gravel or sand is probably near at hand if someone will only wake up and look for it. The second reason is not worth discussing, for an unballasted track is expensive to maintain; much more costly than if well ballasted.

The different kinds of ballast occupy about the following order of merit: broken stone, clean coarse gravel, furnace slag, engine cinder and clean sand. Almost any limestone or granitic rock will form a good ballast, but very soft sandstones and clay or shale rocks should not be used. Soft sandstone breaks in tamping, while the clay or shale rocks, although they may be hard when put into the track, fall to pieces very rapidly when exposed to the weather. Burnt clay has also been tried but, so far as can be learned, it is much inferior to either broken stone or gravel. This is usually because the ballast partakes too much of the character of ordinary building brick and is too little uniform in hardness. Even the best of it is said to fracture quite easily. However, where stone and gravel are practically unobtainable with coal and clay available, burnt clay is probably the best material for that part of the country.

Kinds of ballast.

Broken stone.

It is not by any means an universally accepted opinion that broken stone ballast is better than gravel. Many engineers believe that the greater ease with which gravel may be handled more than compensates for the lasting quality of stone, but it seems almost certain that for roads having many and heavy trains nothing in the end is so good as clean broken stone.

Stone is practically indestructible and almost immovable when once placed and properly tamped. When kept clean it permits of the most perfect drainage, and when it becomes foul, may be cleaned by simply handling it with forks, when it is as good as new. It costs more to prepare a track with stone ballast, but on the other hand costs less to maintain it, while in addition it is probable that the ties last considerably longer in stone than they do in gravel ballast, other circumstances being the same.

There are two general ways of procuring stone ballast, the first of which is for the railroad company to own and operate a quarry and crusher; the other is for the company to buy its broken stone of some contractor delivered on cars. Each way has its special advantages, but it is usually more satisfactory for a road to own and operate its quarry, since it will then have a supply of ballast under all contingencies.

Stone quarries.

Quarries differ so in location that none but the most general description of the best way of operation can be of much use. The crusher should if possible be placed high enough to discharge the ballast into the cars by gravity, and far enough from the loading track to permit of placing a car between the ballast car and the crusher to receive the screenings. The screenings are a valuable by-product of stone ballast and should not under any circumstances be wasted. For certain purposes such as sidewalks and platforms at small stations they are excellent, but they should be excluded from the track since they will impair, to a large extent, the drainage capacity of the ballast.

Conveyors. Although it is desirable to transmit the crushed stone directly from the crusher to the cars by gravity, a good quarry need not be ignored because that cannot be done. The art of conveying by buckets, which are filled and dumped automatically, has reached a high state of perfection, and material may now be transported for long distances, up, down and around corners without the interposition of a single pair of hands and at a small cost of operation.

If it is desired to supplement the supply of stone from the quarry, by stone brought from other points, a track may be carried from the level of the main track up to the top and back of the crusher, by means of which the stone may be conveniently unloaded. This track should gradually descend from the crusher until it meets and connects with the track on which the crushed stone is loaded. The switch connected with this may be automatic and fitted with a spring so that when the car descends from the top of the crusher it will of itself set the switch right, and the spring will replace the switch to that position which will send the car under the crusher on its return journey. The tracks running under the discharge of the crusher should be built on a grade of not less than 50 ft. to the mile so that cars may be moved without the help of an engine.

Size of stone. Stone should be crushed so that it will pass through a $2\frac{1}{2}$ inch ring; anything larger is too large for good track work, and if the stone be broken smaller the percentage of screenings will become too great. Certain flat stones, much larger than this will slip through the crusher, and these, if they are not returned to the crusher and so reduced to the proper dimensions, must be broken by hand after the ballast has been dumped on the track; in any event every section gang using stone ballast should have a supply of napping hammers.

Rotary crushers. Of crushers there are two general types. The "Rotary" (fig. 53a) consists of a heavy

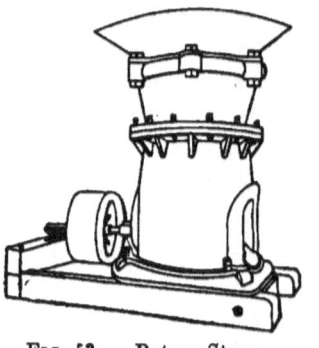
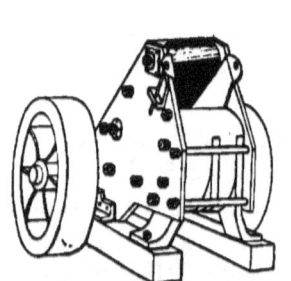

Fig. 53a.—Rotary Stone Crusher. (Gates.)

Fig. 53b.—Jaw Stone Crusher.

cast-iron casing which has a conical opening down through the center. In this opening is supported a powerful, solid, corrugated cone which not only revolves around its own axis but is supported at the bottom by an eccentric which also revolves and forces the cone now near to and then away from the sides of the inner casing. The stone is dumped into the top, and by the eccentric motion of the solid cone is gradually broken as it passes down where, at the bottom, it is discharged from the chute, of the proper size.

Jaw crushers. The "Jaw" crusher is shown in fig. 53b, and operates by means of a fixed plate, which has opposed to it another plate hinged at its upper end and moved at the lower end; this alternately increases and diminishes the opening between the two plates, crushing the stone in the operation.

New embankments. Stone ballast should never be placed upon a new embankment, for it will certainly settle and destroy the established grade; further, if any ballast has been laid, it will be covered by the material which is used to bring the track up to the proper level. The use of stone for this purpose is too expensive, and it is therefore better to wait until the track has settled before stone ballasting is begun. The subgrade should be made of the full width and crowned sufficiently to discharge water freely before any ballast is placed upon it.

Cleaning ballast. Stone ballast should never be handled on the ground with anything but a fork (see illustration in chapter on tools) which will not pick up dirt and can be pushed into the ballast with comparative ease. A screen should also be provided for cleaning the stone whenever it is taken from the ties. This should be done at intervals, usually of about three years, at which periods it will be found that the interstices between the stones are nearly, if not quite filled up with cinders.

Gravel ballast. Although, as has been suggested, stone ballast is the best material for supporting a railroad track for a heavy train service, clean coarse gravel is a thing which every trackman should think himself extremely lucky to get. It is cheaper than stone, easier to handle, easier to raise tracks with, permits of fairly good drainage and, through its use, a good track may be secured,— for a time. Gravel however will not carry the water off so fast as broken stone and it cannot be successfully cleaned except by an expensive process of washing.

Gravel pits. The gravel pit should be located as near the railroad and as near the center of the division to be ballasted as possible. The deeper the cutting in good material, the better, in order

Gravel pits. to avoid moving the shovel oftener than is absolutely necessary, while the pit should be as long as possible so that the switching engine may handle a large number of cars at one time. There should also be plenty of room for switching tracks so that the loading and distributing trains need not be delayed by waiting for each other.

The cheapest way is to buy the land, strip it of gravel and then if possible sell it; but if that plan is not convenient the gravel may be paid for by the yard, purchased for a lump sum to be removed in some specified time or an annual rental may be paid for the land with the privilege of taking out as much or as little gravel as the railroad company pleases. In any case the pit must be carefully stripped before the ballast is removed because nothing is worse in the track than the loam which covers most gravel deposits.

Distributing gravel. If much work is to be done, a steam shovel (figs. 54 and 55) must be used and worked day and night if necessary, for in these days of ballast plows and electric or gasoline lights there is no difficulty in loading and unloading ballast at night. In this way double the track force can be worked with each steam shovel and double the work accomplished with each plant. If the road is poor on which the ballasting is being done, much may be accomplished in having the cars unloaded by way freights. The cable may be left, each time it is used, at the end of the dump, ready for the next day's work, and the unloaded cars may be switched on to some convenient side track to be returned to the pit by the next freight going in that direction. If, however, a great deal of work is to be done, a sufficient number of trains should be assigned to the service, well manned by intelligent trainhands and equipped with good engines, suitable cars and special appliances for unloading the ballast. The engines particularly should be reliable and able to haul a full train without breaking down. It has been a common idea that any old engine, out of repair and almost ready for the scrap-heap, is good enough for maintenance-

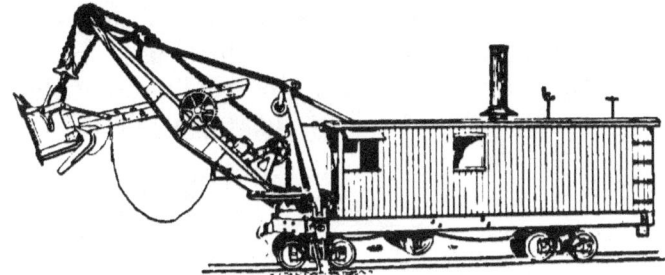

Fig. 54.—Sixty-ton Steam Shovel.
(By the Bucyrus Company.)

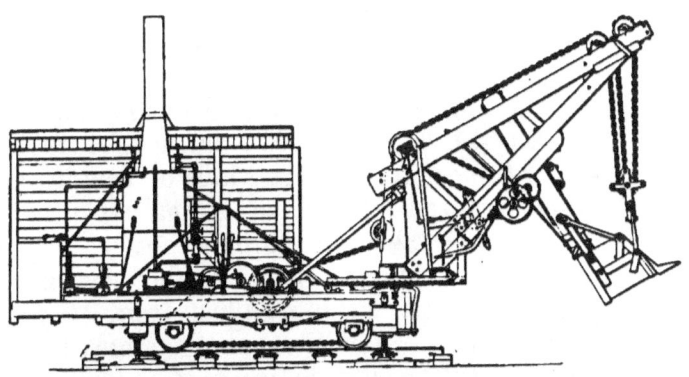

Fig. 55.—A Steam Shovel, showing interior of cabin.

of-way service, but if the cost of the delay that is caused by breaking down at inconvenient times were considered, it is certain that a considerable economy would result from the use of reliable motive power in this service. No simple delay of any train is so costly to a railroad as the delay of a work train.

The steam shovel (figs. 54 and 55) should be **Steam shovel.** able to move itself forward without the use of a rope and should have a long reach. It should be carefully and regularly inspected, and any little damage should be promptly repaired. Six or eight men, besides the engineman, will be necessary in attendance on the shovel, whose duties will be the poling down of the ballast, and the laying of track for the shovel.

Ballast plow. An unloading plow is now a necessity to every railroad, and one which uses stakes on the side of a flat car for guiding the plow as in fig. 56 instead of the old-fashioned center

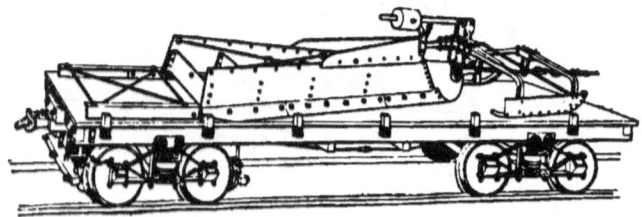

FIG. 56.—Barnhart Ballast Plow.

strip, is the best. The center strip is not only an expense but an inconvenience because it unfits the car for certain kinds of freight, and when required for ballast is often not present. When side stakes only are necessary, any flat car may be taken from its regular service and placed at once in the ballast train. If the brake-wheels are on the ends they must first be changed to the sides in any case.

Unloading methods. Three kinds of unloading arrangements are shown in figs. 57 to 62. In fig. 57 a form of hoisting engine (which however exerts in this case a horizontal pull) is mounted on a flat car and receives its steam directly from the locomotive. It is said to be able to unload any material which could be used for ballast or be taken from a ditch or cutting. From the fact that the unloading device is independent of the locomotive, the plow may be moving in

BALLAST. 83

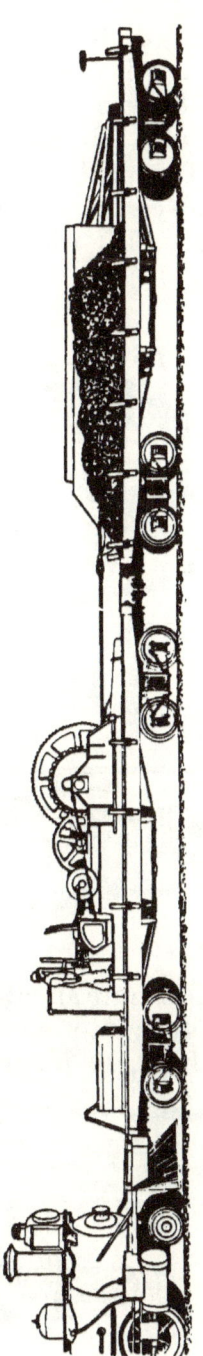

Fig. 57.—Lidgerwood Unloader.

Ballast unloader.

Ballast cars.

one direction while the train is moving in the opposite direction making it possible to have the ballast distributed between the extremes, of all in one place, or in a thin sheet over a long piece of track. The method illustrated in figs. 58 and 59 involves the use of a train of bottom-dump cars (fig. 58) which are followed by a plow (fig. 59) for clearing the track of the ballast.

FIG. 58.—Rodger Ballast Car.

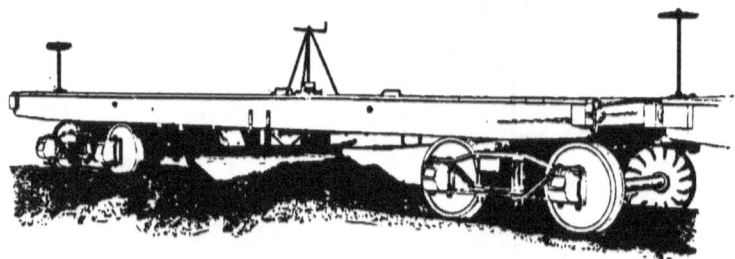

FIG. 59.—Rodger Plow Car.

Figs. 60, 61 and 62 illustrate an all-steel car which dumps in several different ways by means of the inclined sides which are hinged at the

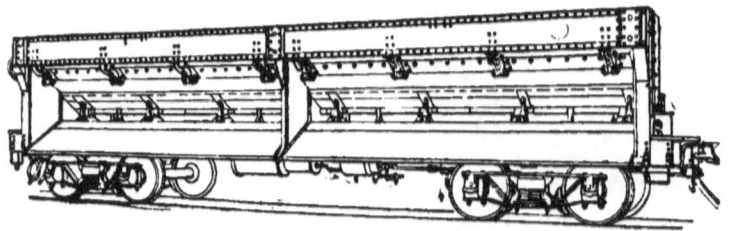

FIG. 60.—Goodwin Dump Car. Side View.

BALLAST. 85

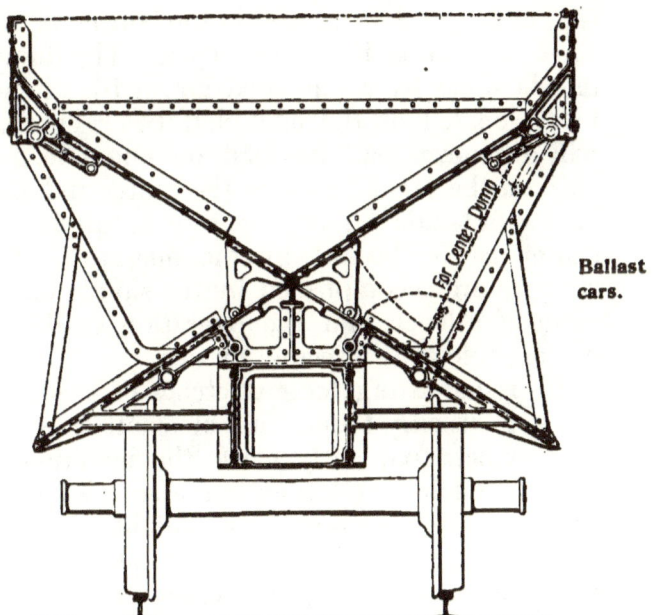

Ballast cars.

FIG. 61.—Goodwin Dump Car. Sectional View.

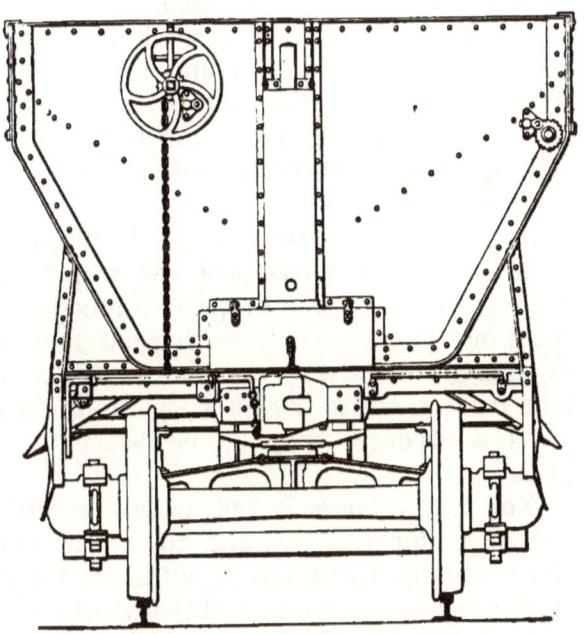

FIG. 62.—Goodwin Dump Car. End View.

Ballast cars. top, and by valves (traps) in the bottom of the car and running its whole length. The dumping is done by compressed air or by a hand-lever located on the end platform. This car will discharge half its load on one side and half on the other; half in the center and half on the outside; all on one side or all in the center as is desired by the operator. The changes are accomplished by the simple movement of a lever, and the operation of dumping occupies but a second or two.

Re-ballasting. In re-ballasting long stretches of road, the track is usually found full of small sags and hills, which frequently cause the breaking in two of trains and largely increase the cost of hauling. The opportunity should then be taken of re-establishing the original, or a better, grade. For this purpose, levels should be run on the rail and after a careful inspection of the profile, grade stakes should be set for the guidance of the trackmen. It is surprising what good results at small cost can be secured by a little care and forethought in this matter. The alinement of the track may also be corrected at this time better than at any other, particularly on bad curves and long tangents.

Heaving spots. In certain spots on nearly all railroads the usual amount of ballast will not stop heaving even when the road is properly drained; these spots or "pockets" may be dug out and filled in with gravel or a special line of tile may be laid from them to the ditch, while in extreme cases a recourse may be necessary to both plans.

Ballast sections. Not less than 8 in. of ballast should be placed under the ties, and the more there is, the better the track will be, although for practical purposes 12 in. is sufficient. An inspection of figs. 63, 64 and 65, which are the bal-

BALLAST. 87

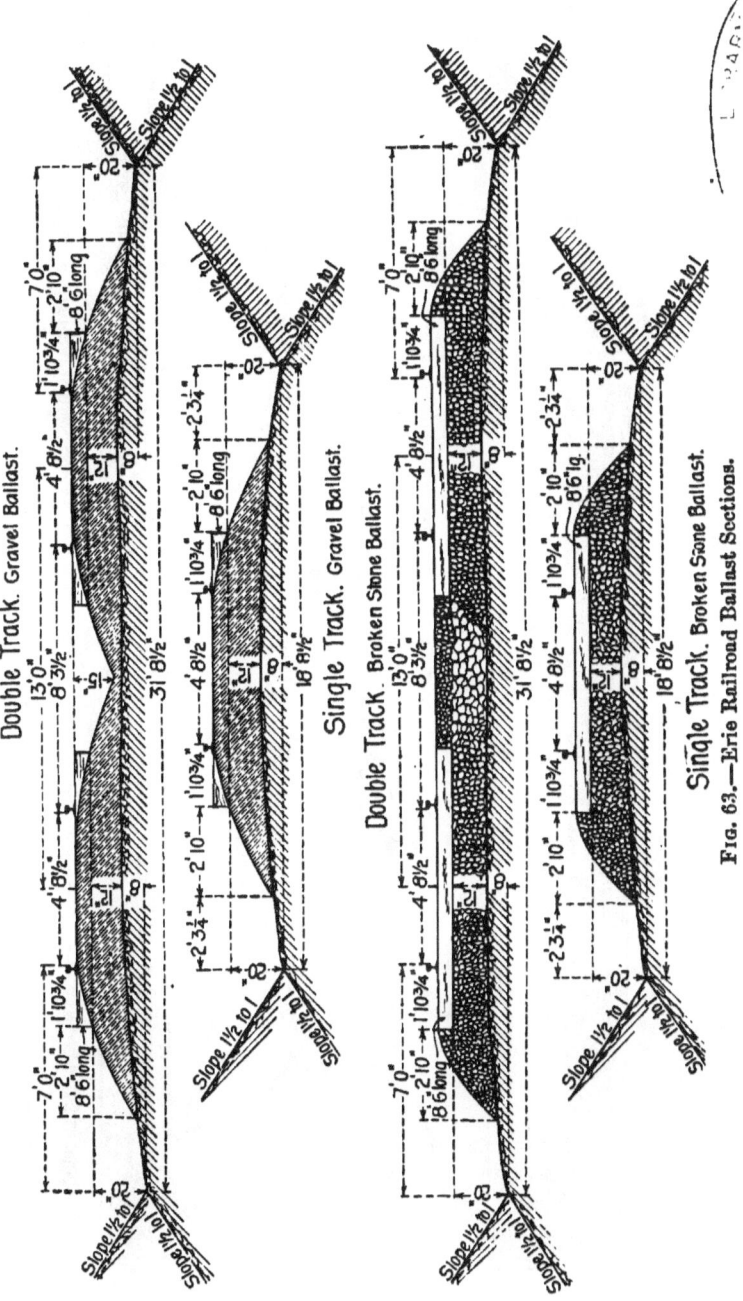

FIG. 63.—Erie Railroad Ballast Sections.

88 THE NEW ROADMASTER'S ASSISTANT.

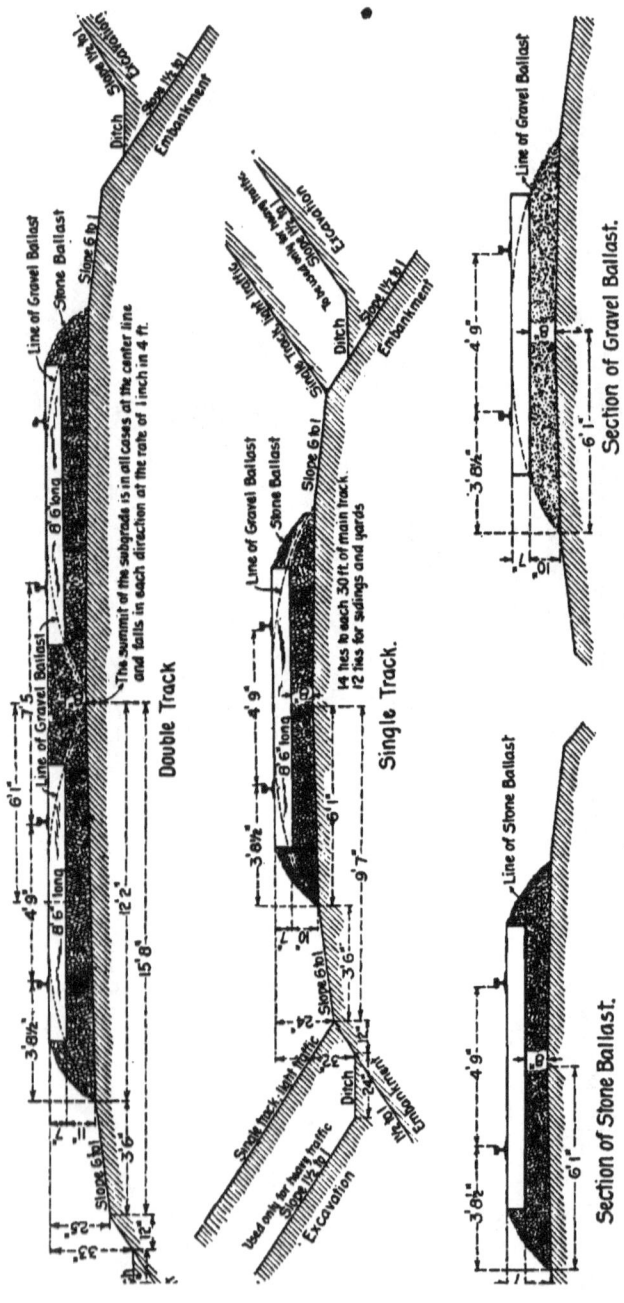

FIG. 64.—Pennsylvania Railroad Ballast Sections.

BALLAST. 89

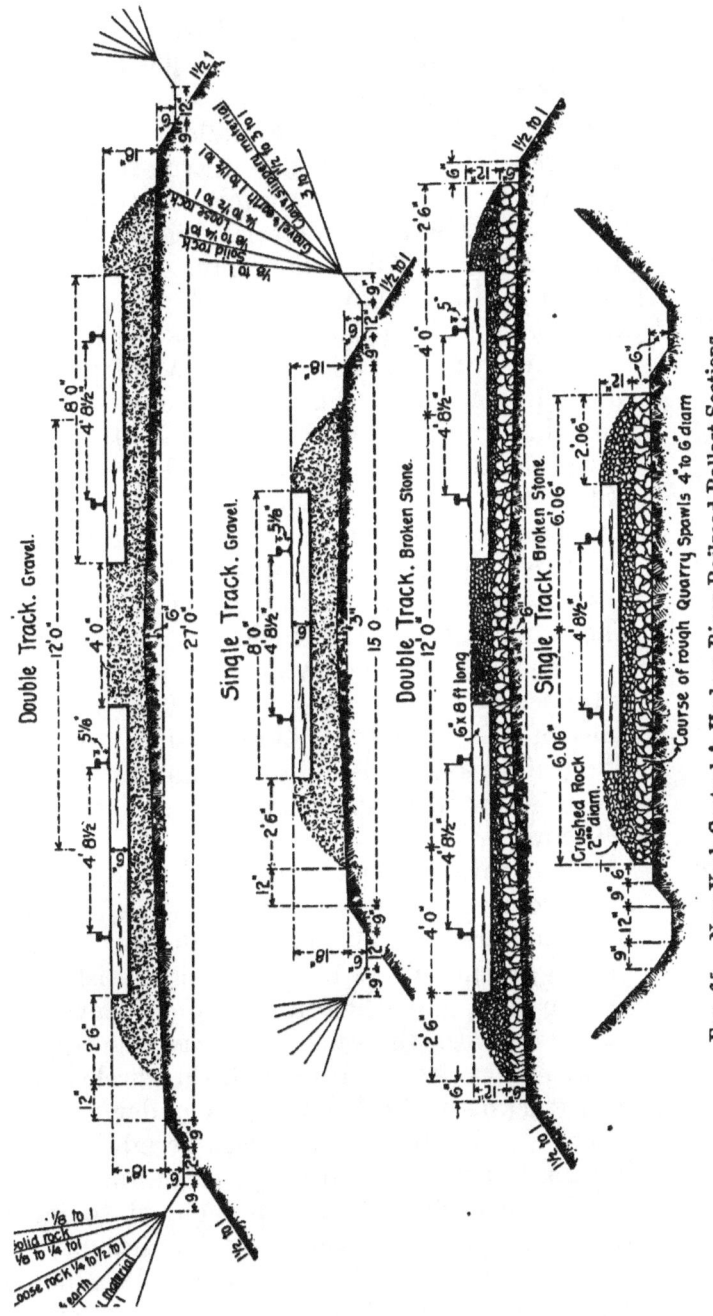

FIG. 65.—New York Central & Hudson River Railroad Ballast Sections.

Ballast sections. last sections of three great railroads, will show the diversity of practice. It is not the intention of the writer to judge between them, but the attention of readers is called to the full and generous lines of the gravel section on fig. 65; these seem to promise a greater stability than where the material begins to slope from the rail or from the center of the track.

Ballast drains. In fig. 66 is shown what is probably the average of all practice as regards the top and side lines of the ballast, wherein the portion of each section on the left represents stone, and that on the right, gravel. In this case, however, the sub-grade of the double track differs from the ordinary types, in that the drainage is made to simulate that of single track by sloping its top at a grade of 25 to 1 in each direction from the center of the ties. The water between the tracks is then collected by a line of 6 in. tile having open joints, which is tapped at right angles every 100 ft. by cross lines of 3-in. tiles with closed joints. Something of this sort is quite common in Europe and might well be used here, where perfection is the aim.

Track jacks. Some form of track jack, of which there are several good ones, should always be used in raising track. The jack should be strong, but light enough to be moved short distances by one man, and when in use should never be placed between the rails. At least one fearful accident has occurred through the carelessness of a trackman who left the jack under the rail where a train struck it and was derailed, causing the injury and death of a large number of people; a sufficiently strong incident to show the results of carelessness on a railroad.

The jacks shown in figs. 67, 68, 69 and 70 are of the prevailing form and are all of well-known make. Figs. 71, 72 and 73 rep-

BALLAST. 91

Ballast sections.

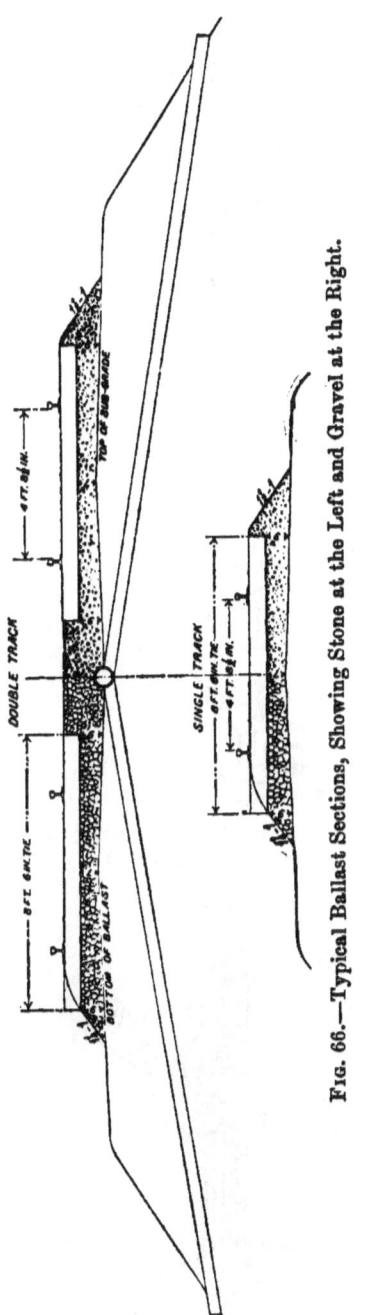

FIG. 66.—Typical Ballast Sections, Showing Stone at the Left and Gravel at the Right.

Track jacks.

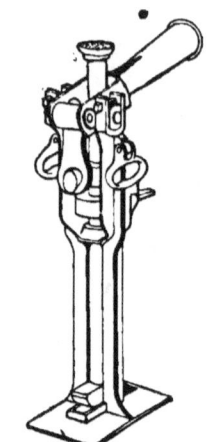

FIG. 67.—Jenne Track Jack.

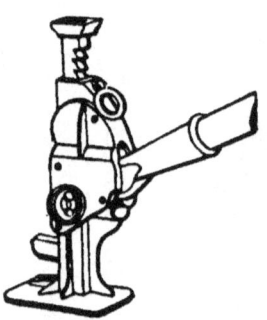

FIG. 68.—Barrett Track and Bridge Jack.
(The Duff Mfg. Co.)

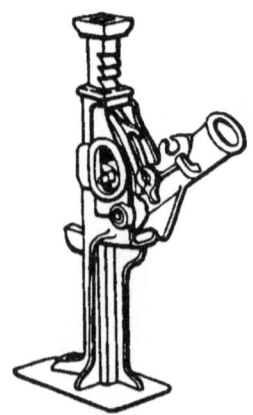

FIG. 69.—Barrett Trip Track Jack.
(The Duff Mfg. Co.)

BALLAST. 93

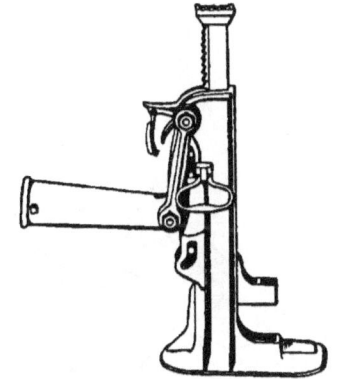

Track jacks.

FIG. 70.—Boyer & Radford Track Jack.

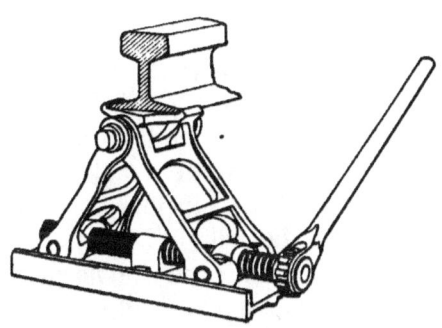

FIG. 71.

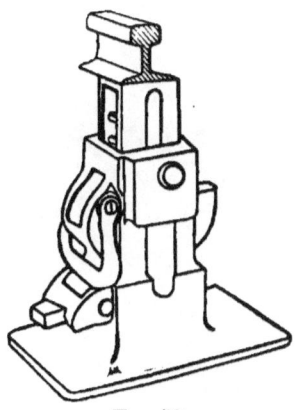

FIG. 72.

Track jacks.

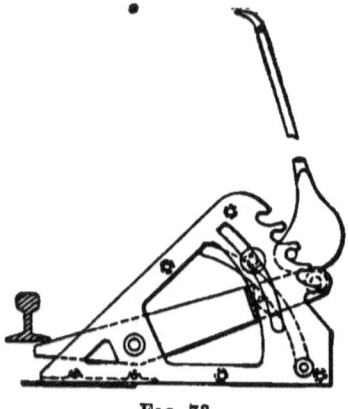

Fig. 73.

Figs. 71, 72 and 73.—Track Jacks for use Under the Rail.

resent the effort to reduce the jack to such a height that none of it will be above the rail. This is laudable so far as it goes, but is objectionable because it may lead to carelessness through the notion of some trackman that a jack is strong enough to support a train.

Raising track.
Great care must be taken in raising track, to have caution signs located or flagmen sent out a sufficient distance from where the work is going on to enable trains to reduce their speed.

CHAPTER IX.

Cross Ties.

Tie Inspection. The proper selection, inspection and distribution of cross ties is one of the important duties of a roadmaster, since in this matter he comes more in contact with those persons who have something to sell than in any other way. All sorts of tricks are practiced to conceal defects in the material; red oak passes as first-class white oak, ties are piled in such a way as to show only those parts which are up to the specifications etc. A roadmaster should therefore be forever on his guard against such possibilities.

Date of inspection. Where ties are bought along the line, a certain time in each month should be taken for inspecting and counting them. This gives the dealers a chance to be present when their ties are counted and enables the roadmaster to settle almost all questions and disputes, (some of which are sure to arise) without wasting time and with very little trouble. All ties should be plainly and indelibly marked on one end with paint or a stamp, when they are inspected, in such a way as to make it impossible for them to be presented a second time, and to facilitate the work of inspection they should be piled in alternate layers at right angles to each other, with a space of 6 or 8 inches between each two ties of the same layer. It is well to distribute the ties as soon as possible after inspection, and dealers should be required to remove all their rejected ties from the right-of-way within a stated time.

Size and quality.

The ordinary dimensions for cross ties are 6 in. x 8 in. x 8½ ft., and they should not be less but may properly be increased to 7 in. x 9 in. x 8½ ft. They must be of perfectly sound material and only one tie should be cut from a section of a tree. This latter practice insures the fact that the timber is young and of second growth. It is best that all ties shall be cleared of bark before they are paid for by the railroad company but in order that the inspection may be reliable, the bark of white oak must be left on until after the inspection. When the ties are shipped from a distance this arrangement may not be convenient but it may nearly always be managed by an agreement with the contractor and should be if it is a possible thing.

Sawed or hewed.

If the ties are hewed, they must be dressed with the faces parallel, for a warped tie will surely make bad track. A prejudice against sawed ties exists which is to a large extent founded on the belief that a sawed tie will not last so long as a hewed tie taken from the same tree. This is probably not true. The reason that sawed ties do not as a rule last so long as hewed ties is that they are frequently cut from large (and that often means old) timber, which has already survived its usefulness. The same objection exists with regard to split ties and on the other hand is impossible in the case of hewed pole ties. Sawed ties, which are surely known to have been cut from young, vigorous timber are perfectly fitted for the main track, while at switches or frogs, on bridges and in all places where the ties are laid upon stringers, no hewed ties should be used since their surfaces are neither flat nor parallel.

It is well, in any case, to purchase a certain proportion of sawed or split ties at a reduced price for use in side-tracks, as they will do quite as well for that purpose, and it will probably tend to lessen the cost of first-class ties by permitting the farmers and dealers to make use of timber which would otherwise prove a loss.

Time for cutting. Where it is possible to require it, ties should be cut in the middle of summer or in winter, at which times the sap is not in motion. The winter is the best time, since then they may be distributed exactly where they are wanted before the spring work begins. By this practice, a time of the year is utilized during which little else can be done and the way is prepared for the rapid placing of the ties in the track as soon as the season permits.

Tie timber. Next to white oak, which combines the qualities of holding a spike and resisting decay longer than any other timber that we have in common use, chestnut, yellow pine and black and red cypress are probably the best, but are closely followed by cedar, tamarac, and in the far western States, redwood.

The supply of white oak fit for cross ties is nearly exhausted in the northern States, and the larger roads now procure many hundreds of thousands a year from Virginia, West Virginian, Kentucky and some of the other southern States.

Tie plates. The probable annual rate of steam and electric railroad building in the United States will draw still further on a rapidly decreasing supply, both for the construction of new roads and their maintenance after they are built. It is therefore of great importance that some means shall be taken to lengthen the life of ties, not only because there is real danger of exhausting the supply but because of the direct economy which will follow their increased service. It may be extended consider-

Tie plates. ably by the use of a tie plate of the type known as the "Servis" (fig. 74), a device which long since passed the experimental

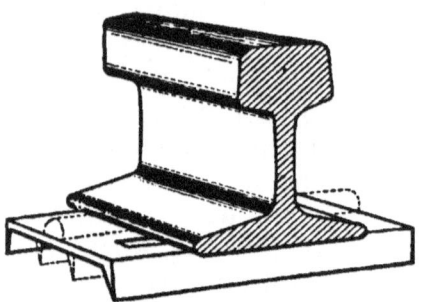

FIG. 74.—Servis Tie Plate.

stage. In this form, the plate is made from a piece of steel whose longitudinal outside edges are sharpened and bent downwards at right angles. Sometimes intermediate ribs and a rail brace are added, but these are not essential and it is quite doubtful if they increase the value of the plate; they are shown as dotted lines in fig. 74. The first object of the tie plate was to preserve the tie from the pounding and crushing effect of the rail, but time developed other qualities of almost equal value. The lateral strength of the track was much increased by the support which the plate offered to the spike. It is of course on curves that this action is most plainly seen, but even on tangents it may be noticed, particularly where track is not maintained in good surface. Simple as it is, this little piece of metal has rendered immense benefits to the railroads of the United States. Tie plates of flat metal had been tried many times but they failed in every case, either because they were too thin and buckled, or because they were too thick and acted as anvils; and in either they were loose and permitted abrasion of ties and spikes. It is the longitudinal ribs which give the plate value.

Some wood-preserving process, such as creosoting, will probably become a necessity and, it is believed by many, would prove an economy even now. There is some uncertainty as to how much the life of a tie would be prolonged by this process, but from European and American experience it is probable that instead of having to renew all of the ties in the main track every seven or eight years, it would be necessary to do so only once in every fifteen years or even more. The saving resulting from this would be greater than is apparent, not only in the price of the ties, but in all the labor of inspecting, distributing and of putting them in, as well as from having to disturb the track so infrequently, a reason which every trackman will appreciate. In Europe, where timber is much scarcer than here, creosoted ties have been for many years in common use and are giving, on the whole, the best results.*

Wood preserving process.

Iron and steel ties have also been used extensively in Europe, and in this country to a very limited extent. There are many forms of metal ties, each of which claims to have special advantages over all others, and several of which resemble each other closely, so that if their use is decided upon a choice should not be difficult. There are a few situations, as in busy tunnels, where it is possible that their use is warranted even at their present cost, for they may be expected to last as long as the rails.

Metal ties.

In fig. 75 are shown two typical forms of metal ties. Each of them is made from a

---

* Bulletin No. 9, of the Forestry Division, United States Department of Agriculture, gives an account of this question together with that of tie plates and of metal ties which should be read by every maintenance-of-way officer. It can be procured by an application to the Secretary of Agriculture, Washington,

Metal ties. thin sheet of iron or steel, rolled or pressed into shape, and the rail is held to them usually by a clip, bolt and nut, as in fig. 76, although as

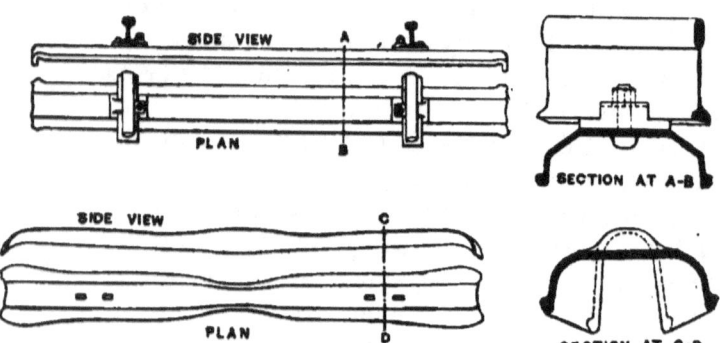

FIG. 75.—Two Forms of Metal Ties.

explained below, the sheets of fibre which surround the rail are only used under exceptional conditions.

Tie Insulation. Where metal ties are used in conjunction with the electric track circuit (see Chapters IV

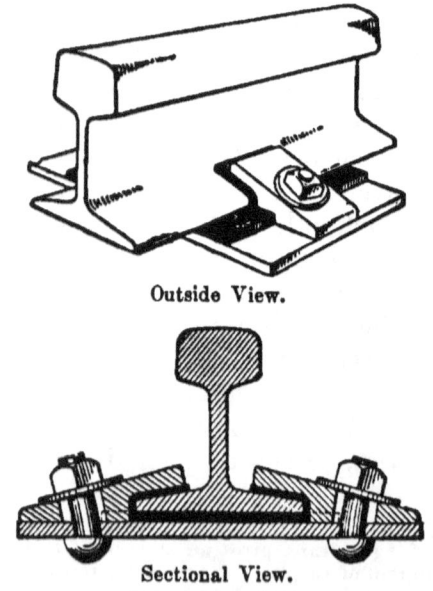

FIG. 76.—Method of Fastening the Rail to a Metal Tie, with a Fibre Insulation.

and XV), it is necessary to prevent the electric current from passing from one rail to another by means of the cross ties. This can only be done by some method of insulation. Fig. 76 illustrates one of the methods. Here there are some fibrous sheets interposed between the rail and all of the surrounding metal; since the fibre is a non-conductor of electricity, the rail becomes what is called "insulated." *Tie Insulation.*

A certain side of the track should be fixed upon for lining the ties, if the road is single track. On a double track road the ties should be lined always on the outside of the track except on curves where they should always be lined on the inside of the curves. *Lining ties.*

The track jack should be used as much as possible for putting in ties, since it saves the labor of a man who would otherwise be engaged in holding up a tie for another man to tamp. It should be particularly provoking to an ambitious roadmaster to go over his track and see men on every section lazily roosting on a crowbar. *Track jacks.*

Late in the fall the roadmaster should, in the company of each section foreman, make a careful examination of his track for the purpose of ascertaining the number, kind and location of the ties which must be renewed during the next year. When this location is known the new ties can be distributed at any time during the winter. *Counting ties.*

Unfortunately, ties in any section of track do not all decay together. The practice of removing all at once, quite desirable on some accounts, may easily result in serious waste; and in this connection it should be remembered that ties made from different kinds of wood should not be mixed in the track, since they will decay in different lengths of time. *Entire removal.*

Spacing ties.

On main track, from fourteen to sixteen ties, depending on their size, should be laid under a thirty-foot rail, and on side track from eleven to thirteen ties. If the ties are large there will be less than this number, and if they are small there will be more. They should be evenly spaced with reference to their bearing edges (not their centers) and sufficient room for tamping should be left between each two ties. For joint ties in main track this distance should be from eight to ten inches, not more, and for intermediate ties from ten to fourteen inches. The largest ties should be reserved for placing under the joints.

Tamping ties.

Ties should be tamped hard at the ends and for about a foot inside the rails, but in the middle less tamping should be done; just enough to solidify the earth and hold the end tamping in place. The shocks to which a tie is subjected, all come under the rail at which point the tie will first settle. If the tie is tamped hard in the middle, it, in time, becomes center-bound and rocks, which is bad for the track, and disagreeable to the passengers. Neither should ties be tamped too high at the joints in the expectation that they will gradually settle down to the proper level of the track. A joint which is too high is as injurious to the rail, and has as bad an effect on the train as a low joint.

Time to renew.

As has been said in a previous chapter, the work of putting in ties should be commenced in the spring as soon as the most important of the ditching has been completed, and ought, if it is a possible thing, to be finished by the first of July. This is a rule which cannot be too strongly insisted upon.

Disposal of decayed ties.

After the decayed ties have been taken from the track they should be disposed of promptly and not be left to disfigure the right-of-way.

They may be burned on the spot, taken to the **Disposal of decayed ties.** nearest engine-house to be used for starting fires in the locomotives, or the people living in the neighborhood may be permitted to haul them away, but before they are disposed of in any way they must have been carefully inspected by the roadmaster. This should never be omitted, since it is the only possible means of checking the care with which the section foremen select the ties for removal. It is of course best to utilize them if possible, and there are many ways in which to do it. They make good temporary retaining walls when placed ends into the bank, and if laid side by side across very muddy roads form a good substitute for "corduroy." If no better use presents itself, they can be piled around stumps and burned, thus "killing two birds with one stone." In any event, they should not be left scattered along the track but should be gathered into neat piles at the close of each day's work.

In yards, in the neighborhood of switches, **Use of pegs.** or where the track heaves, cross ties are often rapidly destroyed by the frequent drawing and re-driving of spikes which, each time it is done, leaves an opening for the water to settle in and exposes the center of the tie to decay. To prevent this, a supply of pegs a little larger than a spike, should be kept on hand for filling these holes and should be used invariably where a spike is drawn from a tie which is to remain in the track. The pegs are best made by machinery, but if the supply should become exhausted they may be split out of good timber at the hand-car house on rainy days. The practice of driving a pick into a tie on any and every occasion for testing or moving it is a bad one. It makes a hole in which water will settle and start a center of decay. If it is

ever really necessary to pick a tie, a sound peg should be driven tightly into the hole and cut, not broken off at the surface.

Shims. Shimming is a makeshift at the best and is apt to be greatly overdone, often to a dangerous extent. It is almost entirely due to bad drainage, and to correct this is the only proper course. If, however, circumstances require shims let them be made and used in the right way. Hard wood is the only suitable material from which to make shims, white oak preferably, and they should be sawed and bored by machinery; they then are of the best form and cost the least money. Those more than an inch and less than two inches in thickness should have four holes, two for the rail spikes and two for spiking the shim to the tie. All shims more than two inches thick should have the ends beveled with two holes bored in each end for spiking them to the tie. For this purpose, long boat-spikes should be used.

Spiking ties. In spiking ties the arrangement of the spikes should be invariably as shown on fig. 121, with both inside spikes on the same side of the tie and with both outside spikes on the other side of the tie. If the spiking is done in any other way the tie will twist instead of remaining always at right angles to the rail. Although apparently a small matter this is really of much importance. Driving spikes properly requires some skill and is more often badly than well done. The gage should always be used and the spike stood up straight beside the rail and touching it, then driven down straight, not leaned in either direction to suit the convenience of the trackman.

Level and gage. Any work which tends to disturb the ballast of track should be performed with the assistance of a track level, and the work of putting in ties should prove no exception to this rule. **Track levels.** The level should be moved from place to place where the men are tamping and no track should be considered finished until it has been finally tested with the level and gage and is known to be right.

CHAPTER X.
Rails and Fastenings.

Many of the details of track have been greatly modified during the last few years, particularly as to the form of the rail and to those parts which have to do with holding the rails together. The simple fish-plate has practically disappeared except for side tracks; the six-hole angle bar is driving out the four-hole angle-bar, and some form of trussed or bridge joint may take the place of the six-hole angle-bar. The iron rails are already a thing of the past, except for scrap, and at the present time if any railroad company could be found to purchase them they would cost more than those made from the best steel. *Recent changes.*

The form of the rail itself has undergone a radical change during the last ten years, brought about by a great many, and at the time, largely unaccountable failures, which attracted the attention of railroad engineers, rail makers and inspectors, and resulted in a careful and accurate investigation into the causes of the trouble. *Early rail defects.*

Speaking in general terms, this investigation showed that the form of the rail was largely at fault. Through defects in design, the rail makers were forced to finish it too hot, resulting in a rail the head of which had been insufficiently worked and, what was still worse, received its last working at a high heat. The same faulty design caused the rail to cool quicker in the head than in the flange, which made it difficult to finish it straight, and gave rise to internal strains.

Rail sections. The present most approved form is illustrated below in fig. 77, and is the section of an eighty-pound rail now in use on some of the most important railroads in the country. This

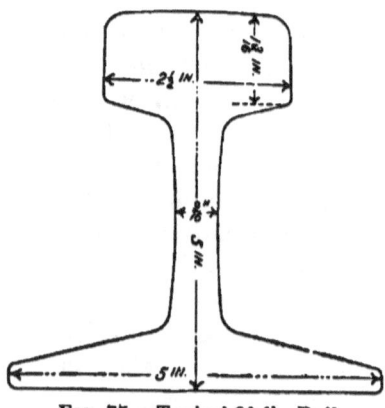

FIG. 77.—Typical 80-lb. Rail.

section has been developed from the labors of P. H. Dudley, long identified with the improvement in track on the New York Central & Hudson River, Boston & Albany, and other railroads, together with the efforts and investigations of Messrs. Robert W. Hunt, J. D. Hawks and D. J. Whittemore, aided by several others, members of the American Society of Civil Engineers. This society has now put its stamp of approval upon the general form of rail urged by these gentlemen, and there seems little likelihood of a return to the old hit-or-miss practice.

Comparisons. The aim of the new section is to secure the best possible distribution of the metal, in order to correct the troubles which experience has developed. For the purpose of comparison an eighty-pound rail (fig. 78), designed in 1887, is illustrated, and in fig. 79 the two first sections are superposed; they show quite plainly the difference between what was commonly considered a very good section in 1887 and that which conforms closely to the best present practice. The most important difference is in the general form, which has been

Rail sections.

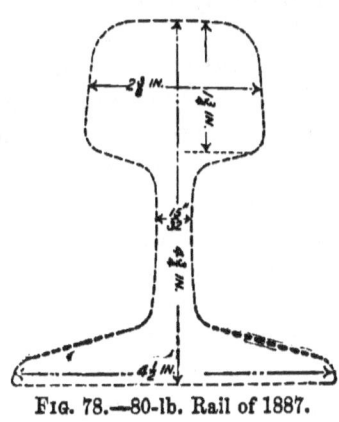

Fig. 78.—80-lb. Rail of 1887.

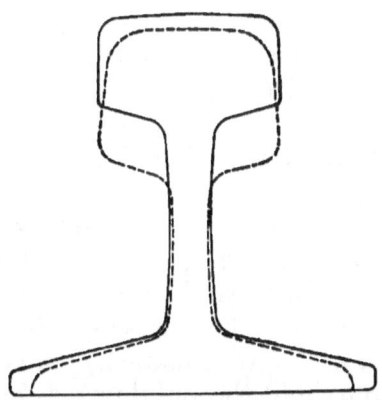

Fig. 79.—Comparison of Figs. 77 and 78.

made higher in proportion to weight; the web has been thickened and increased in length while the head is much thinner with both base and head wider than formerly. The increase in the height of the rail has greatly strengthened it as a girder, and the increased width in the head has provided a larger surface for carrying the heavy modern rolling stock which rapidly destroys the earlier forms of rail by literally squeezing the metal from the top of the narrower heads.

Composition. Rails are now being made much harder than was usual in the past, since instead of containing but 0.25 or 0.30 per cent. of carbon, they often have 0.55 or 0.60 per cent. and the tendency is to still increase. It is not decided as to what is just the proper quantity of carbon in order to secure the most efficient combination of hardness, toughness and elasticity, for it must be understood that, speaking in a general way, the greater the percentage of hardening elements, the more brittle the metal.

Rolling. It will be seen by a comparison of figs. 77 and 78, that the relative quantities of metal in the head, the web and the base of each section are much more nearly equal in the former than in the latter. This has rendered it possible to roll the rails at a much lower temperature than could be done when such forms as fig. 78 prevailed. As a consequence the density, toughness and reliability of the metal have been greatly increased.

Spikes. There are many forms of spikes, varying slightly from each other, but no one form seems to have gone far toward superseding the old-fashioned spike. The lag screw (fig. 80) (in the opinion of many trackmen), in connection with a clip to hold the rail down, as in fig. 112, is likely to supersede the driven spike, and other persons just as well informed regard it already as a demonstrated failure. Fig. 81 is the same as fig. 82, but twisted and with a sharp point. Fig. 82 is the common $\frac{9}{16}$-in. by $5\frac{1}{2}$-in. spike; figs. 83 and 84 are modifications of fig. 82, while fig. 85 is quite different since it is intended to be driven at an angle

across the tie but parallel with the base of **Spikes.**
the rail.

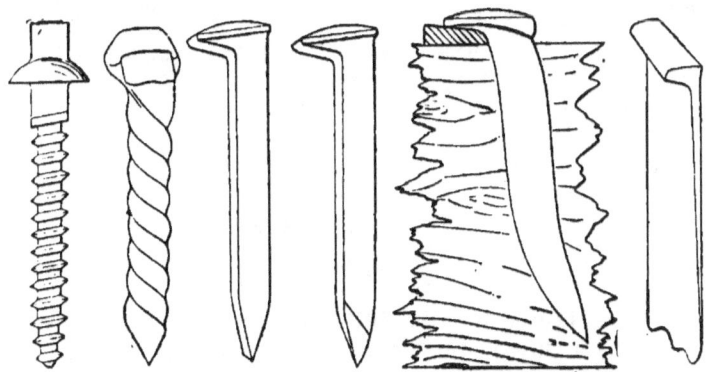

FIG. 80.   FIG. 81.   FIG. 82.   FIG. 83.   FIG. 84.   FIG. 85.

FIGS. 80–85.—Different Forms of Lag-screws and Spikes.

There are also large numbers of nut locks, **Nut locks.** most of which have followed the well known "Verona" idea. The original and the latest Verona patterns are shown in figs. 86 and 87, while figs. 88, 89, 90, 91 and 92 illustrate variations of it. The first Verona nut lock (fig. 86) was formed of a square steel rod, bent into a circle, whose ends were cut diagonally and then forced a little apart. It acts in two ways: When the nut is screwed up tight the nut lock acts as a spring, forcing the

FIG. 86.—Original Verona.   FIG. 87.—Recent Verona.

FIG. 88.—Eureka.

FIG. 89.—American.

Nut locks.

Fig. 90.—Harvey.

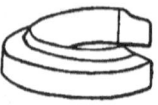

Fig. 91.—National

Fig. 92.—Excelsior.

Spring Nut Locks.

threads of the nut and bolt together, while its sharp ends form cutting edges which bear respectively against the splice-bar and the nut and oppose the effort of the nut to revolve. In its new shape, one of the ends is prolonged, which when it strikes a projection, acts as a stop to prevent the nut lock from turning. An entirely different arrangement is presented in fig. 93. Here the washer is put on flat as in the upper figure, and as shown by the dotted lines of the lower figure. The nut is then screwed home, when the pressure forces the small projections in the center hole against the thread of the bolt. The tongues are then

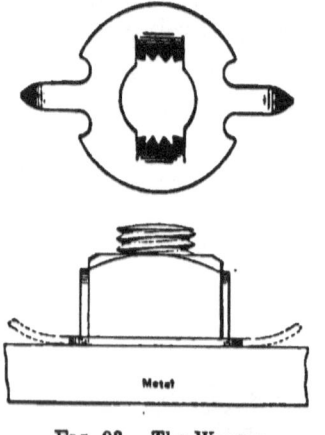

Fig. 93.—The Warren Washer for Metal.

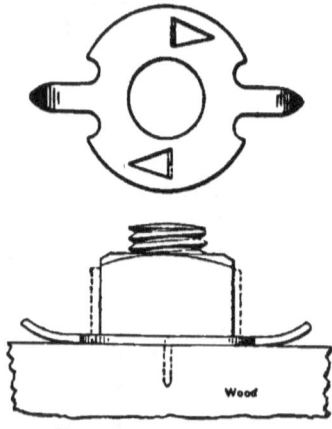

Fig. 94.—The Warren Washer for Wood.

turned up against the side of the nut. Fig. 94 **Nut locks.** is of the same order, but is intended for wood and may therefore be used on timber culverts, trestles and bridges. Although not of the same class as the previously described nut locks, fig. 95 comes under the same category.

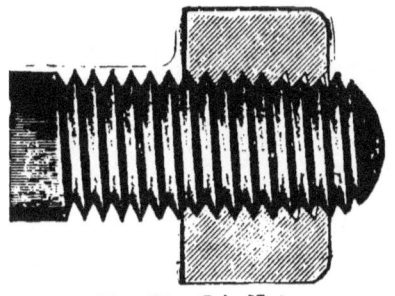

FIG. 95.—Grip Nut.
(Oliver Iron & Steel Company.)

It differs from the ordinary nut only in having the three outside threads made at a little different angle from the others and this causes the nut to bind against the bolt.

Of joint fastenings there is a great variety, **Rail joints.** many of which will be found illustrated in figs. 96 to 105. Theoretically, the rail at the joint, in order to secure a perfect track, should be exactly as flexible and as strong as the rest of the rail. So far, however, the fault has been chiefly that the joints have been too weak. The requisites in a joint are, in the order of their importance, strength, ease of application, fewness of parts and cheapness. Cheapness is put last because a poor joint at a less price, will prove more costly than a higher priced good joint. Fewness of parts is a very desirable feature since the more parts a joint has, the more there are to become loose and get lost and the longer it takes to put them together. Ease of application is also requisite because the operation of renew-

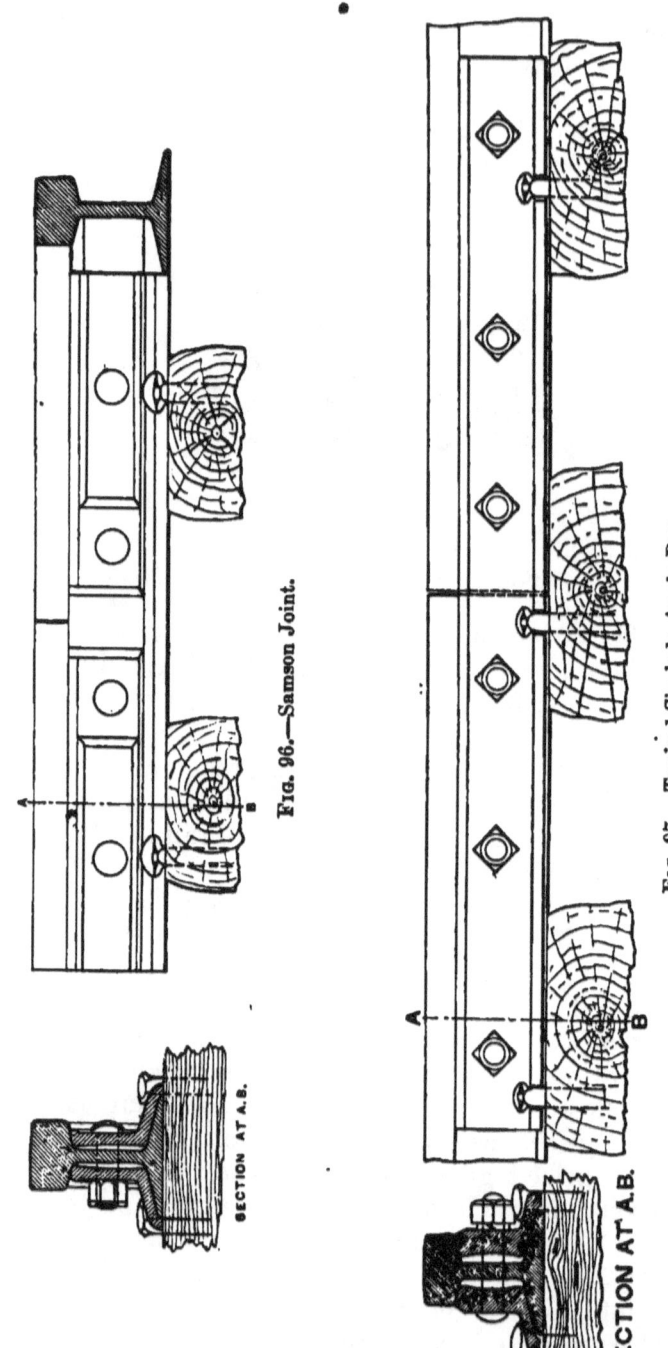

Fig. 96.—Samson Joint.

Fig. 97.—Typical Six-hole Angle Bar.

RAILS AND FASTENINGS. 115

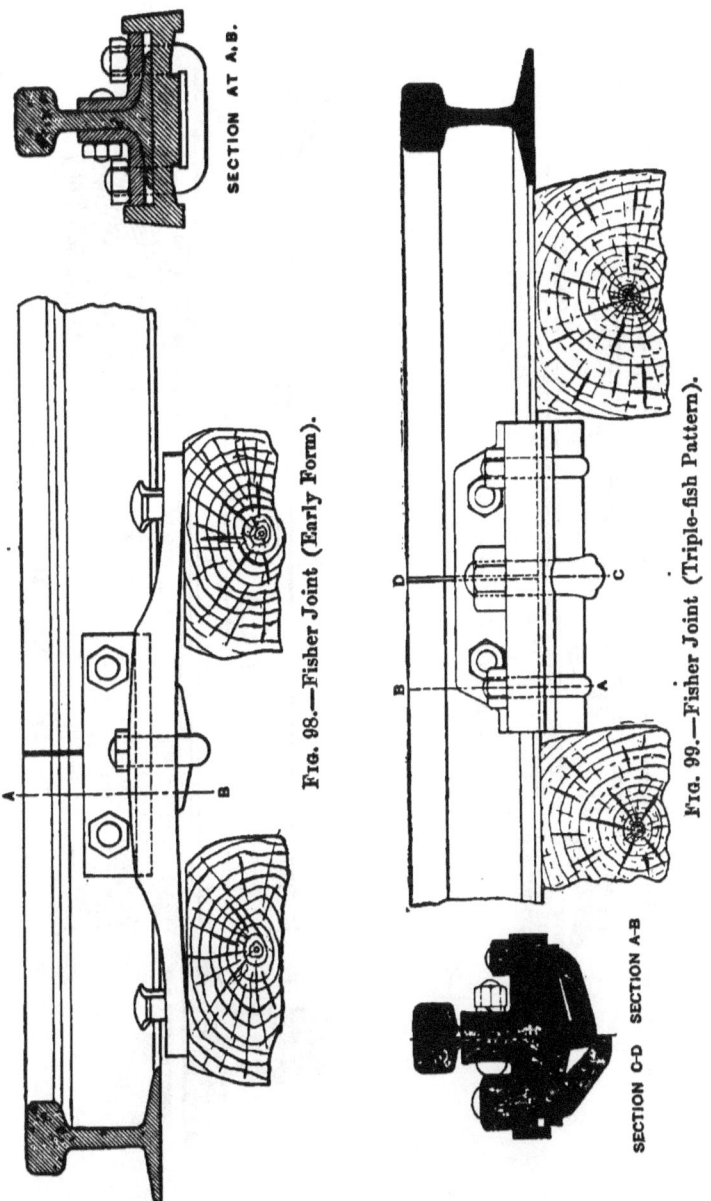

Fig. 98.—Fisher Joint (Early Form).

Fig. 99.—Fisher Joint (Triple-fish Pattern).

116   The New Roadmaster's Assistant.

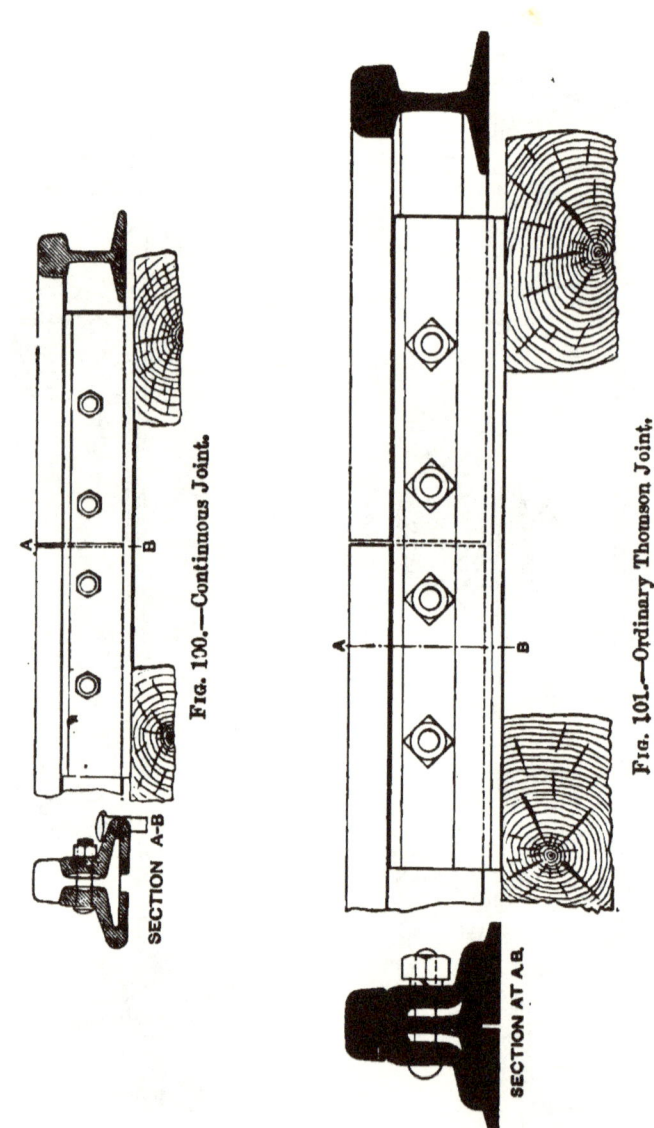

Fig. 100.—Continuous Joint.

Fig. 101.—Ordinary Thomson Joint.

RAILS AND FASTENINGS. 117

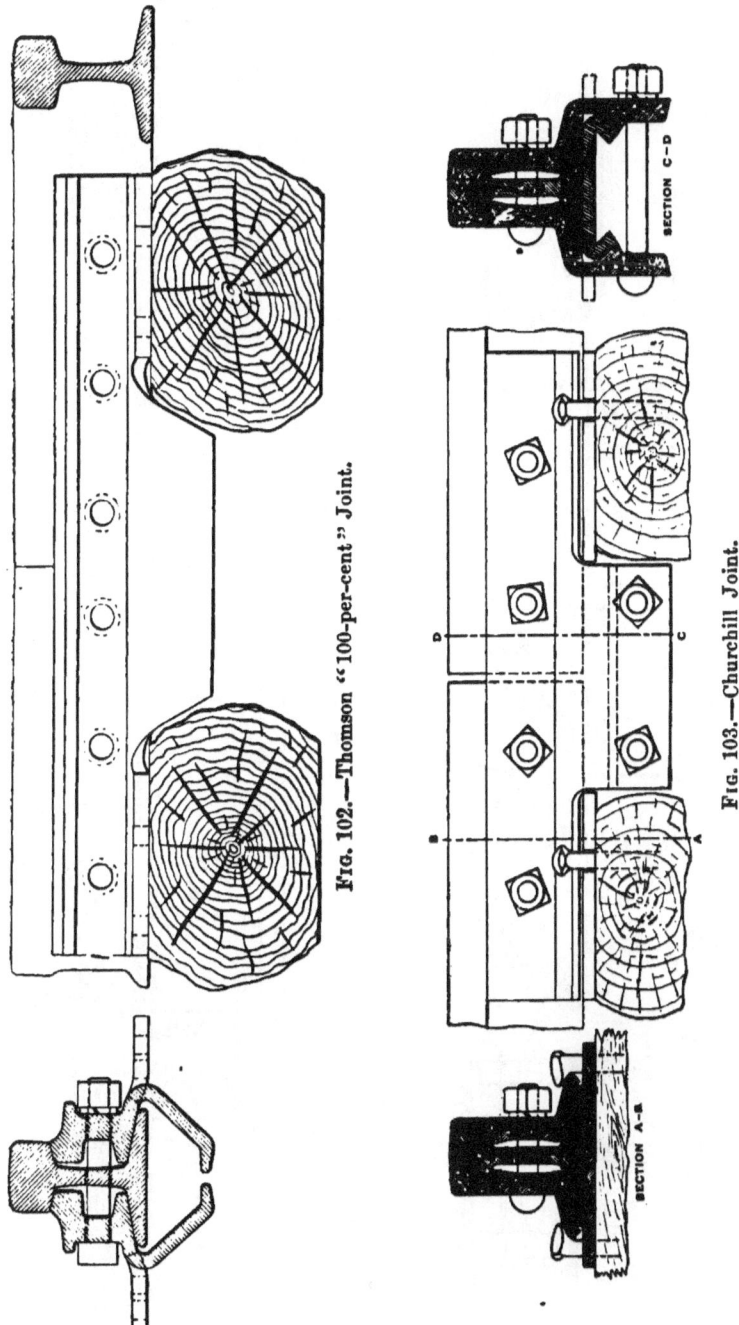

Fig. 102.—Thomson "100-per-cent" Joint.

Fig. 103.—Churchill Joint.

118   THE NEW ROADMASTER'S ASSISTANT.

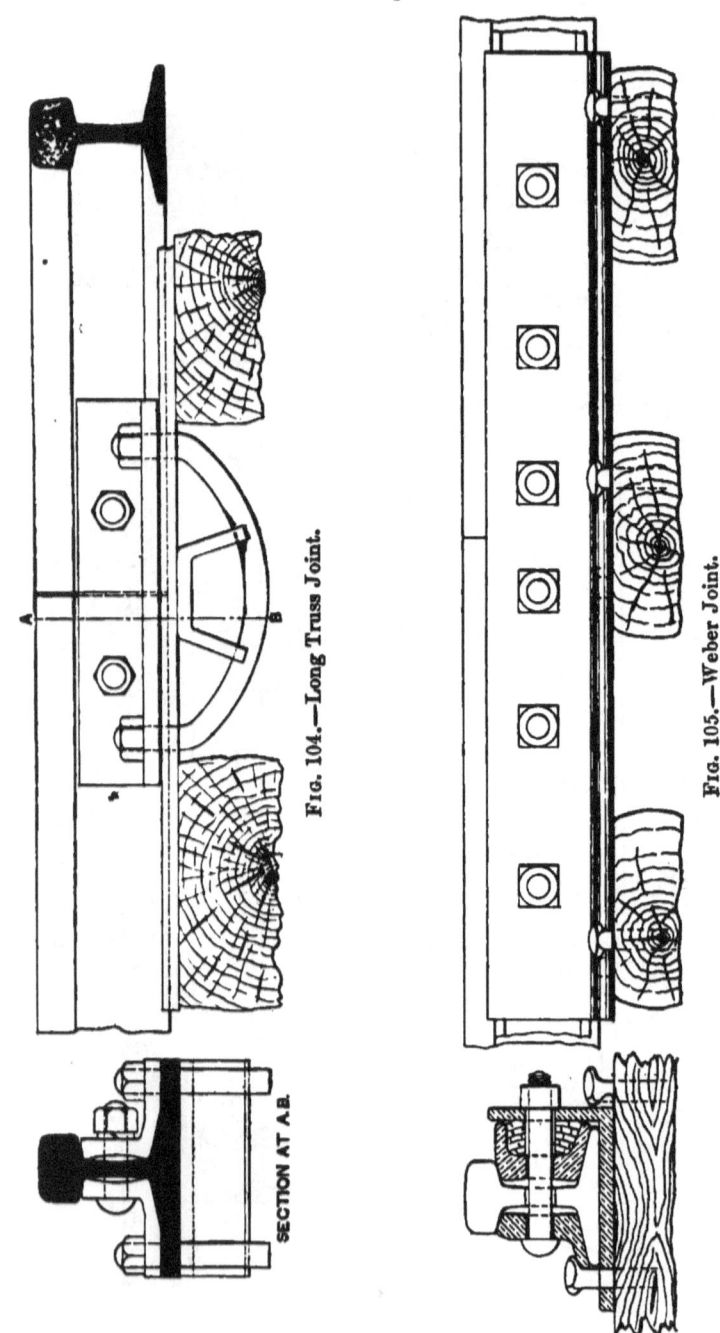

FIG. 104.—Long Truss Joint.

SECTION AT A.B.

FIG. 105.—Weber Joint.

ing rails is usually done in a great hurry and at a time when every moment counts.

It would be impossible in the limits of this book to give so much space as would be required to illustrate a tithe of the various joint fastenings which are even now being experimented with. A large proportion of them are foredoomed failures while in the case of many others experience seems to show that they must also end in that railroad limbo, the scrap heap, there to remain until they are re-melted or re-rolled into some form of usefulness. *Angle bars.*

Fig. 96 shows a four-bolt angle joint, although it is not quite a typical joint except at the section A-B, for in the center the bar has been thickened. This accomplished its purpose, that of preventing the tearing apart of the bar at the joint, but it did not prevent the bar from bending and remaining bent permanently. The six-hole angle-bar is probably much better but there is almost a certainty that it is not good enough and that it cannot be made so. Fig. 97 illustrates a fair example both as to its length and the distance between the bolts. These angle-bars vary considerably in total length, from 32 in. to 42 in., and in extreme cases even less and more. A comparatively small proportion are used as suspended joints (that is with two ties) but this defeats one of their chief advantages and practically places them in the class of four-hole angle-bars. This is a fact which seems to have escaped many people since it appears quite evident that the useful effect must diminish in proportion as the bolts are further from the junction between the rails unless, as with the three-tie joint, the bars are re-inforced by the resistance of another tie. It is widely believed that no spike-slot should be placed at *Suspended and supported joints.*

or very near the center, since this is the point of greatest strain and it is here that most (substantially all) of the fractures occur. One of the best informed and most careful investigators of track materials has stated that "the angle-bar should be high in carbon and low in phosphorus, so that it may be very stiff and elastic. The mild steel splice takes a 'set' after which it holds the rail-ends down, causing a permanent low joint which cannot be corrected until new splices are put in."

Description of joints.

Fig. 98 is, like fig. 96, of historical interest, since it shows one of the earliest attempts at a "bridge joint," in other words the use of two cross ties to sustain the shock of the wheel-blow and the inherent weakness at the break between the two rails. Fig. 99 is the descendant of fig. 98, but neither in form or idea does it resemble its ancestor. It is exactly what its name indicates, since it provides a vertical as well as a horizontal "fishing," and to all intents and purposes, rejects the assistance of the ties.

Fig. 100, it will be seen, is like fig. 96, but with its flanges widened and turned under to act as a support to the bottom of the rail. The great resemblance between figs. 100 and 101 needs nothing more than mention.

In fig. 102 we find one of the most recent joints. The splicing parts are exceptionally heavy and the wide flanges are bent down between the ties to act as a further opponent to the up and down movement of the rail.

Something of the ideas expressed in figs. 98 and 102 are found in fig. 103, but with the addition of bolts to assist in the labor performed by the downward bent flanges of 102. The "Long" truss joint (fig. 104) differs but slightly from fig. 103, since it is intended to overcome the same strains in practically the same way.

RAILS AND FASTENINGS. 121

Fig. 105 is the six-hole angle-bar of fig. 97, plus the bottom support of figs. 98, 100, 101, 103 and 104, plus a wooden filling-piece and a cover-plate, to protect the wood and equalize the pressure of the nuts.

Besides the changes which have taken place in the joint fastenings, it has been suggested to change the form of the joint itself. Four methods are shown in figs. 106 to 112. The intention is the same in all of these suggestions, that is, to transfer the load gradually from one rail to the next instead of all at once as is done where the rail has a square end. **Various rail-ends.**

FIG. 106.—Mitred Rail-end.

Fig. 106 shows what is called the "mitred end," a plan that has been largely followed on the Lehigh Valley Railroad.

FIG. 107.—Vertically-Halved Rail.

In fig. 107 is seen a method which contemplates rolling the rail in halves. This was first tried with iron rails which broke down rapidly but since then, a rail after this fashion has been used in Germany with apparent success. The plan followed, differed from the original one, since with it the rails were permanently joined through the web by frequent rivets. It proved expensive, however, and it is believed that the plan next to be described will be found cheaper and better in every way.

122 THE NEW ROADMASTER'S ASSISTANT.

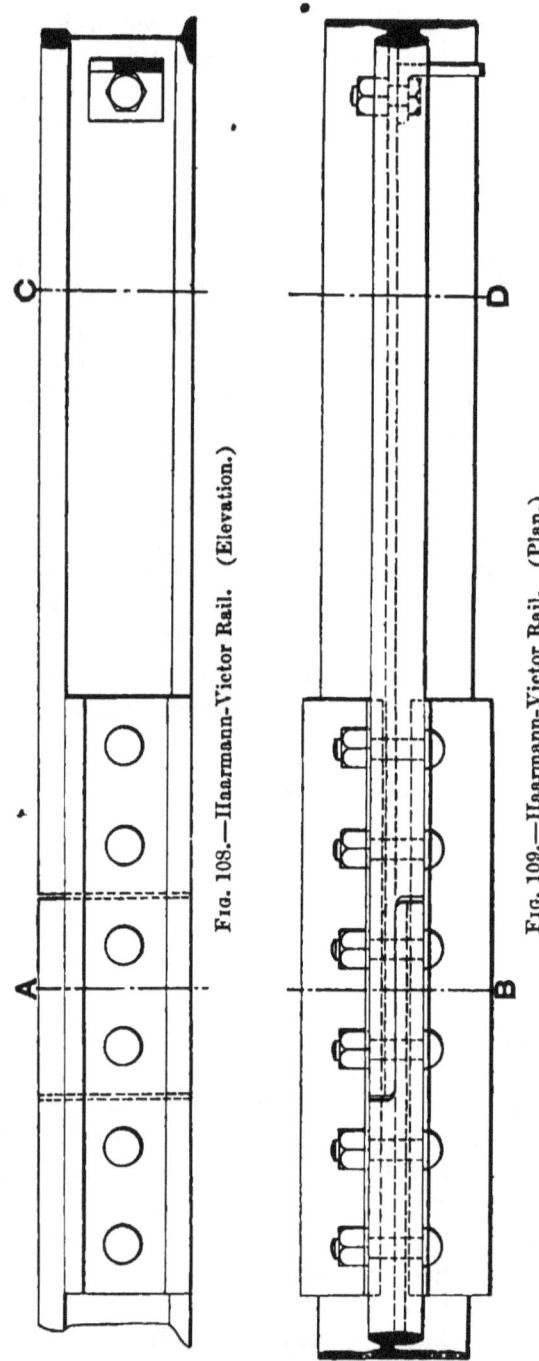

FIG. 108.—Haarmann-Victor Rail. (Elevation.)

FIG. 109.—Haarmann-Victor Rail. (Plan.)

RAILS AND FASTENINGS. 123

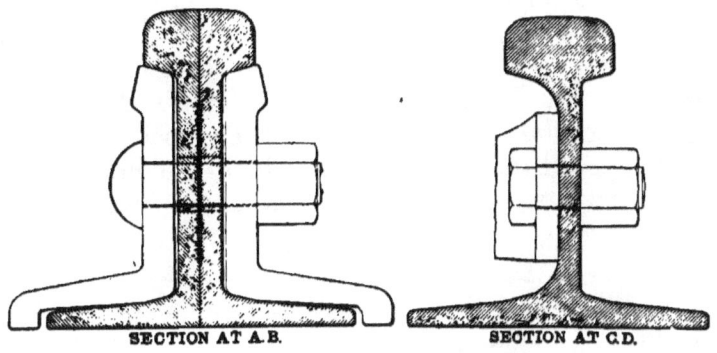

FIG. 110.—Haarmann-Victor Rail. Section at Joint.

FIG. 111.—Haarmann-Victor Rail. Section at Middle.

Figs. 108, 109, 110 and 111 illustrate that system of track known as the "Haarmann-Victor," named for its designers, two German engineers of reputation. As will be seen from fig. 111, a rail of irregular shape is used, which permits that one-half the head and base shall be cut away for about ten inches at each end without removing any of the web of the rail in the operation. The webs are then overlapped and further reinforced by angle-bars as shown in figs. 108, 109 and 110, while the two lines of rails are joined at frequent intervals by tie-straps, shown in different positions on figs. 108, 109 and 111. This rail is very large. It has a height and base of about eight inches each, and is embedded in the ballast. A peculiarity of the plan is that the rails rest directly upon the ballast without the use of any cross ties. The reports of its performance state that a record of ten years' service prove that the cost of maintenance in Germany did not exceed twenty dollars per mile per year, while the stiffness of the splice and the continuity of support secure an unparalleled smoothness of track. It is evident that nothing but the best construction throughout and the most

Haarmann-Victor system.

thorough drainage can be used in this method of laying track, which seems to have given good results in experimental service.

Katté plan. Mr. Katté, chief engineer of the New York Central & Hudson River Railroad, has proposed the use of a rail which is composed of two parts (fig. 112), a head and a base. These

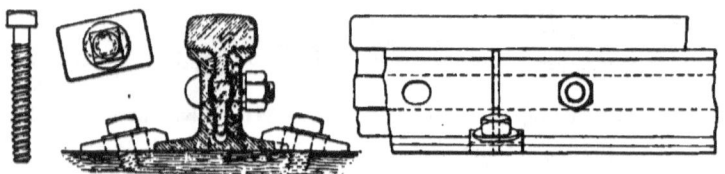

FIG. 112.—Katté Rail. Combined Rail and Splice.

portions overlap each other one-half, and they evidently dispose pretty thoroughly of the "joint question," since the angle-bars (or rather what corresponds to them) have in this case become continuous.

Longer rails. Right in line with the attempts which are being made to secure perfectly tight and rigid joints are the comparatively recent experiments in laying 45 ft., 60 ft. and even 90 ft. rails. This has already been largely done and proved to be a decided advantage; an easily understood matter when it is remembered that 45, 60 and 90 ft. rails have respectively but $\frac{2}{3}$, $\frac{1}{2}$ and $\frac{1}{3}$ as many joints as a 30 ft. rail.

Continuous rails. Practically continuous rails have been successfully used in this country. The idea, which was patented on one occasion by a section-foreman named Noonan, seems to have been first tried on what is now a branch of the Norfolk & Western Railway. The rails were bolted together as tightly as possible and their ends were butted, allowing no room for expansion. Roughly speaking, steel expands one inch in every 100 feet for an increase of

100 degrees Fahrenheit in temperature. Until recently it has been believed that the rail should be able to move through the joint or it might buckle, destroy the gage of the track and cause a derailment, but in the light of some experiments made on the Michigan Central Railroad by Mr. Torrey, its chief engineer, some doubt exists as to whether it would not be better to combine the movement of the several joints at particular points considerable distances apart, in the form of special expansion joints. The experiments indicate that the movement of the rail is not so great as might be expected under the changes of temperature which take place, and after four years of service Mr. Torrey still calls a stretch of track which has riveted rails 500 ft. long the best piece of track on the Michigan Central Railroad. The method followed has been to insert split points in the track some hundreds of feet apart, in the manner suggested in Chapter XI for taking up the movement in creeping track and to rivet the intermediate rails ends together in such a way as to make a practically continuous rail.

Extensive arrangements have also been designed for welding or casting the rail ends together. The first method and that which promises the best results is an electric welding plant, mounted on a car, which heats the adjacent ends of rails and melts them together by means of the electric current, thus forming one piece out of several rails. The other plan contemplates the use of a portable cupola or furnace for melting iron, and a series of molds which are clamped around the joints of the rails. The melted iron is then poured into the molds which are allowed to cool and are then removed. There is not much doubt that there is a strong tendency to eliminate at

Casting and welding ends.

least a part of the joints in our railroad track by some of the methods indicated, and it must be counted as a powerful factor in the progress of the immediate future. Not only is this to be seen on steam railroads experimentally, but street railroads have been and are being built, wholly without regard to the expansion and contraction of the rails due to changes in temperature.

CHAPTER XI.

Track Work.

The rails are usually received from the mill on flat or gondola cars and should be carefully counted at the time they are unloaded for the purpose of detecting any error in the shipping list. The brand is always on the same side of the rail, as it lies in the car. Since many railroads do and all railroads should require the rails to be laid with the brands either all on the inside or all on the outside of the rail when in the track, it must be seen to that the cars point in the right direction when the rail is taken out to be unloaded. If the turning of the cars (when they point in the wrong direction) is not done, each rail must be separately turned in order to get the brand on the proper side. In any event, a string of rails should always have the brand on the same side, because the top and bottom surfaces of the rail are not always parallel, owing to an imperfect adjustment of the rolls in the mill; therefore, if the rails are not laid uniformly a rough track is apt to result. *Counting and turning rails.*

A great deal has been written and said concerning the proper way to unload rails in order that they shall sustain the least possible damage in the process. The dangers likely to occur are the breaking and kinking of the rail. The breaking of a rail in unloading may be caused by carelessness, but on the other hand may result in the detection of a flaw which would have rendered it unfit for use in the track. The dropping of a rail on its end or among a lot of other rails is very likely to break or *Unloading rails.*

Unloading rails. bend it in such a way as to render it practically useless, but when it is squarely dropped on the bare ground, the fact that it breaks is sufficient proof that it should never have gone into the main track.

Among other methods which have been devised for unloading rails, there are three which seem to offer some advantages. The first contemplates the use of two drag ropes with a hook at each end. Each rope is handled by a gang of men, who attach it to the track at one end (by placing the hook under the rail) and insert the hook at the other end of the rope in one of the bolt holes of a rail on the car. The train is moved slowly forward, the rope dragging the rail from the car, while the men prevent it from dropping suddenly by receiving it on hand-spikes before the car has quite passed from under it. It is then carefully lowered to the ground. The two gangs alternate in their movements, one gang being engaged in lowering a rail to the ground, while the other is attaching its rope to a rail on the car. Of course it is only possible to use this method when the rails are loaded on flat cars, or on cars with openings in the ends. The second plan (fig. 113) provides a pair of hangers, of

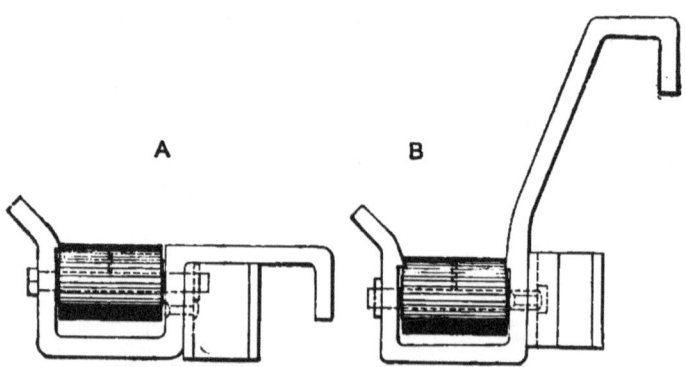

Fig. 113.—Rollers for Hanging to the Side of a Gondola.

different heights, in which rollers are mounted. When the hangers are hooked over the side of a gondola car, the rail is placed in them by one gang of men on the car, and by reason of its slope, the rail slides easily to the ground where it is received by another gang. This arrangement is evidently suited to any car without a roof and also permits, as does the first method, that the rails shall be unloaded without injury while the train is in motion. A third method (fig. 114) is much like the

Unloading rails.

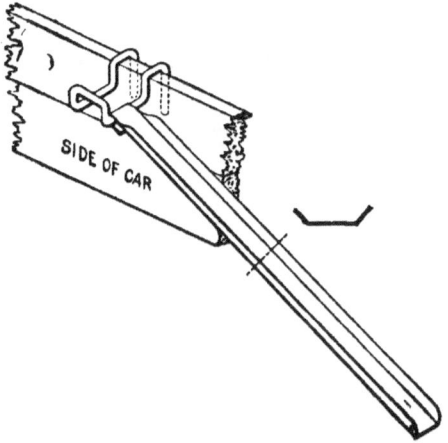

FIG. 114.—Rail-chute to be Hung to the Side of a Gondola.

preceding, except that a chute is used instead of the two hanging rollers. Neither is it necessary to provide a gang to lower the rail to the ground, since the upper end is supported by the chute at all times during the passage.

The skidding of rails is hardly possible under ordinary circumstances, unless time is of no importance.

In re-laying rails it is necessary that every preparation shall be made to facilitate the work before breaking the main track. The rails should be laid end to end, at least partially bolted, and the spacing should be carefully

Re-laying rails.

looked after with regard to the joints on the other side of the track.

Spacing for expansion. In separating the rail ends from each other, iron shims must be used which should vary in thickness from an eighth to three-eighths of an inch. During cold weather the three-eighths inch shim should be used and the one-fourth inch for average temperatures, but in ordinary summer weather an eighth of an inch is quite sufficient, while on days when the thermometer goes above 90 degrees Fahrenheit no opening at all is necessary. The practice of using pieces of wood for this spacing ought not to be permitted. Fig. 115 illustrates a convenient im-

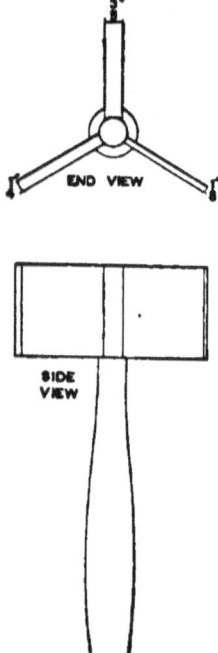

FIG. 115.—Shimming Tool for Laying Rails.

plement, which can be made in any blacksmith shop, where the three thicknesses are provided in the shape of arms set around a stem.

For a temporary connection with the old rail a split point should be kept on hand, which can be bolted to the end of the new rail in a moment and spiked up against the old rail or *vice versa*, when it is desired to let some train go by, or to close up for the night.

Temporary connection.

Where trains are close together the work of tearing out the old rail and laying the new rail should be carried on at the same time, so as to make the most of what interval there is between trains. In this way and by keeping the split point at the head of the gang, the track can be kept open until the last minute.

If the new and old rails are of different heights, "offset splices," (of which three kinds are illustrated in figs. 116, 117 and 118),

Offset splices.

FIG. 116.—Hawks' Offset Splice.

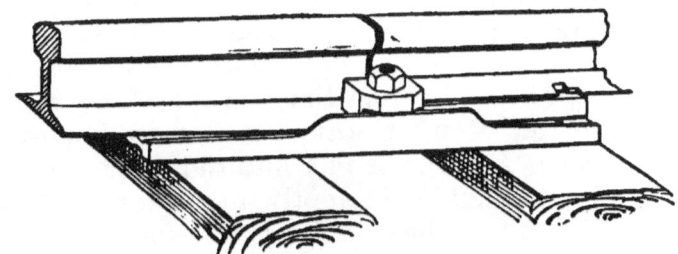

FIG. 117.—Fisher Offset Splice.

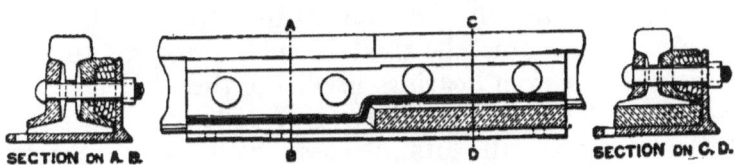

FIG. 118.—Weber Offset Splice.

should be provided for the joint at which the connection is to be made.

When the new rail has a wider base than that which it replaces, the ties must be prepared to receive it, for the old rail will have cut in to a certain extent, leaving a shoulder which must be trimmed down before the new rail is laid. This can be done by special men provided with adzes, who will follow the gang which is throwing out the old rail.

Speed of trains.
In preparing for the removal of the old rail most of the inside spikes and some of the outside spikes may be drawn and trains may still be permitted to pass at a low rate of speed; but care should be taken to locate caution signals or men with green flags, sufficiently far from the work which is going on, to warn approaching trains that the track is not safe for high speed, while, if there is any doubt as to the safety of the track even for low speed, red, flags must be sent out.

Time to re-lay.
As has been stated, one of the best periods for relaying rail is that which immediately precedes the opening of spring work. This, however, cannot always be done in actual practice, since rail must usually be laid when it is received, or very shortly after, and this will be at different times in the year. But no matter at what season, it is most important, when new rail is put into the track, that the joints shall be promptly attended to. Where old rail has been in for some time the ends are apt to be bent down and the ties not level, consequently, where all of the ties cannot immediately be raised to support the rail, shims must be used in such a way as to give it an even bearing on all the ties.

Short pieces.
The ordinary length of rails is 30 feet, but among all lots of new rails some shorter pieces will be found which are very useful in

maintaining the proper distance between joints on the opposite sides of the track, while going around curves, or past frogs and switches. By a judicious use of these short pieces almost all cutting of rails may be avoided.

The bolts, nuts, nut-locks, etc., necessary for re-laying rail, should be carefully distributed immediately before they are needed and should not be thrown around helter-skelter to be lost or buried in the ballast. Men will not be careful of material which does not belong to them unless they are closely watched, and this is a matter which should never be lost sight of in doing any kind of track work. Care of material.

The question of opposite or broken joints has now been generally decided in favor of the latter. The prevailing practice on almost every important road in the country is in favor of this, since there is good reason for believing that trains ride more easily over them and with less noise than they do over opposite joints; it is probable too that the track is more easily maintained. Opposite and broken joints.

The increased height in the rail has made it possible to place the nuts on the inside of the track without fear of having them cut off by deeply worn tires. This enables a track-walker to see at a glance all of the nuts on a single piece of track as he patrols his section, without being forced to step from one side of the track to the other, or to patrol different sides of the track at different times. Track bolts should be regularly tightened, for when only slightly loose they do not unscrew nearly so rapidly as when loose enough to receive a considerable motion from the passing of trains, that is, the looser they are the more rapidily they shake off. With reference to their form it is only necessary to state that square nuts are prefer- Track bolts.

able to hexagon nuts since they do not wear out at the corners.

Elevation on curves. The amount of super-elevation which shall be put in the outer rail on curves is a difficult matter to settle and it is probable that no practicable rule, which can be applied to every case, will ever be arrived at. The three elements in every question are, the general character of the traffic, the maximum speed of trains and the degree of curve. If it were not for the first it would be possible to compromise the requirements of the last two elements very satisfactorily. But it is evident that to arrange a track for a speed of sixty miles an hour, around curves of five or six degrees, would be inadvisable if there were but one or two trains to use the track at that speed. A happy mean is therefore the nearest that can usually be arrived at and the simplest is also the best rule for this purpose. Three-fourths inch for each degree of curve with a maximum of six inches is easy enough to remember and is a safe rule. The result would be as follows:

SUPER-ELEVATION FOR A SPEED OF 40 MILES PER HOUR.

Degree of Curve,	$\frac{1}{2}$	1	$1\frac{1}{2}$	2	$2\frac{1}{2}$	3	$3\frac{1}{2}$	4	$4\frac{1}{2}$	5	$5\frac{1}{2}$	6	$6\frac{1}{2}$	7	$7\frac{1}{2}$	8
Elevation in Inches,	$\frac{3}{8}$	$\frac{3}{4}$	$1\frac{1}{8}$	$1\frac{1}{2}$	$1\frac{7}{8}$	$2\frac{1}{4}$	$2\frac{5}{8}$	3	$3\frac{3}{8}$	$3\frac{3}{4}$	$4\frac{1}{8}$	$4\frac{1}{2}$	$4\frac{7}{8}$	$5\frac{1}{4}$	$5\frac{5}{8}$	6

If the rail is to be elevated for still higher speeds, the rate of increase may be put at 1 inch per degree of curve, that is 1 inch for a 1° curve, 2 inches for a 2° curve and so forth. It would not be safe to advise a higher elevation than six inches for any curve except under rare conditions. If the rail, ties and ballast are respectively as heavy, sound and deep as the best service demands, then and only then is it advisable to exceed six inches of super-elevation. Eight inches is in any case the maximum.

But after all the true way to test the riding quality of a curve is to ride around it. If, on entering a curve, the engine gives a back-breaking twist or the rear car slams against the outer rail in a crack-the-whip fashion, it is a pretty good argument that something is wrong even if the curve has been put up according to an authorized rule.

Where the curve is not "eased," as will be explained later on, the elevation on the curve must be carried back on the tangent and the commonest distance is 100 feet for each inch of elevation. This however would not be possible at reverse curves where there is little or no connecting tangent, in which case the rate must be shortened to suit the conditions. *Tapering off.*

Trackmen are seldom required to arrange for the super-elevation on bridges since the work requires more accuracy than the force is able to command with the ordinary tools of the section. In consequence, special ties, sawed to the necessary taper are usually provided; or else substantial shims, of the same length as the ties, also sawed to the required taper are placed between the rail and tie and spiked or bolted to the latter. If the shims are used, they should be at least an inch wider than the tie, in order that they shall overhang a half-inch on each side and prevent moisture from penetrating and resting between the shim and the tie. The ordinary shim used on heaving track should never be placed under the outer rail of the curve on a bridge. This is because the two rails of the track will not then lie in the same plane and the wheel-treads will consequently roll on the edges of the rails instead of directly on the tops. *Elevation on bridges.*

Curve easements.

The only satisfactory and proper method which is applicable to all commencements of curvature, is "easement." This is done by inserting what is called a "transition curve" between the tangent and the main curve. The transition curve begins at the straight line with no curvature whatever and gradually increases in sharpness until, when it joins the main curve, the two have the same rate of curvature. By this plan, it is possible to begin the super-elevation at the point where the tangent joins the transition curve, gradually increasing the super-elevation as the curve becomes sharper.*

Widening gage.

It is sometimes necessary to widen the gage of the track ; this happens frequently at switches and on side tracks which are used by consolidation or other locomotives which have a long wheel-base. Two inches is the most that the track should be widened and when this is done, guard rails should be placed close inside the outer rail of the curve and close outside the inner rail, in order that the blind drivers may not run off the track. There are many things which would qualify any rule for guidance in this matter, but the principal one is the width of driving wheels, which may vary between five and a half and six and a half inches, so that what is perfectly feasible on one railroad may be impossible on another, owing to the peculiarity of its motive power. Although it is hardly necessary to widen the gage on main tracks (owing to the comparative slightness of the curvature) it is nevertheless the practice on some lines. A fairly average rate for this purpose would be $\frac{1}{16}$ inch per degree of curve, increasing by jumps of two

---

* One of the simplest methods of laying out transition curves is explained in Torrey's "Switch Lay-Outs and Curve Easements," published by the Railroad Gazette.

degrees. That is ⅛ inch for a 2 degree curve, ¼ inch for a 4 degree curve, ⅜ inch for a 6 degree curve, etc.

Rail braces. Rail braces, figs. 119 and 120, must be used on the outside of both rails at curves of more

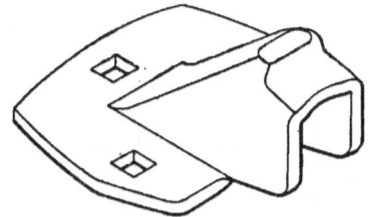

FIG. 119.—Weir Frog Company's Rail Brace.

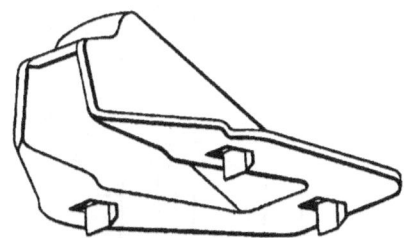

FIG. 120.—Elliot Frog & Switch Company's Rail Brace.

than three degrees, and at all other places where the track is likely to spread. On the easy curves three braces to the rail is enough, but these must be increased as the curvature becomes sharper until there are two braces on every tie. They may be made of die-formed steel as are the figs. 119 and 120, which is the best plan, or of malleable iron or of wrought iron, but a cast-iron rail brace is likely to be worse than useless and should never be used at an important place.

Creeping rails. Creeping rails are the source of much annoyance and sometimes cause serious damage when not promptly and regularly attended to. They occur under different conditions and require different remedies which are often determined by the local circumstances sur-

Creeping rails.

rounding the trouble, but on bridges where there is a heavy grade the fault is most likely to be found and has been entirely corrected by filling the space between the cross ties under the rail with short blocks of oak of the thickness of the ties and spiking the rail closely to them as in fig. 121. The common

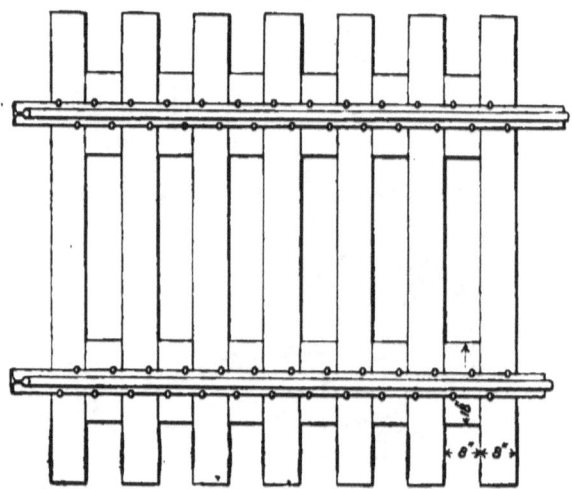

FIG. 121.—Filling-blocks on Bridge Floors.

practice of spiking in the slot holes or at the ends of the angle bars to prevent rails from creeping on bridges is bad and, while it does not often cure the trouble, the ties, if they are not damaged, will usually be disturbed.

On the ground the joints may be braced to the ties by straps of iron, and, if all other plans fail, single split points should be placed in the track, for they may move back and forth a distance of several inches and do no harm. When they have moved the allowed distance, the rails back of them can be changed by inserting pieces of different lengths which must be kept on hand for the purpose. An ingenious and useful application of this idea is shown in fig. 122. Here the split rail is unspiked but is held firmly against the main

TRACK WORK. 139

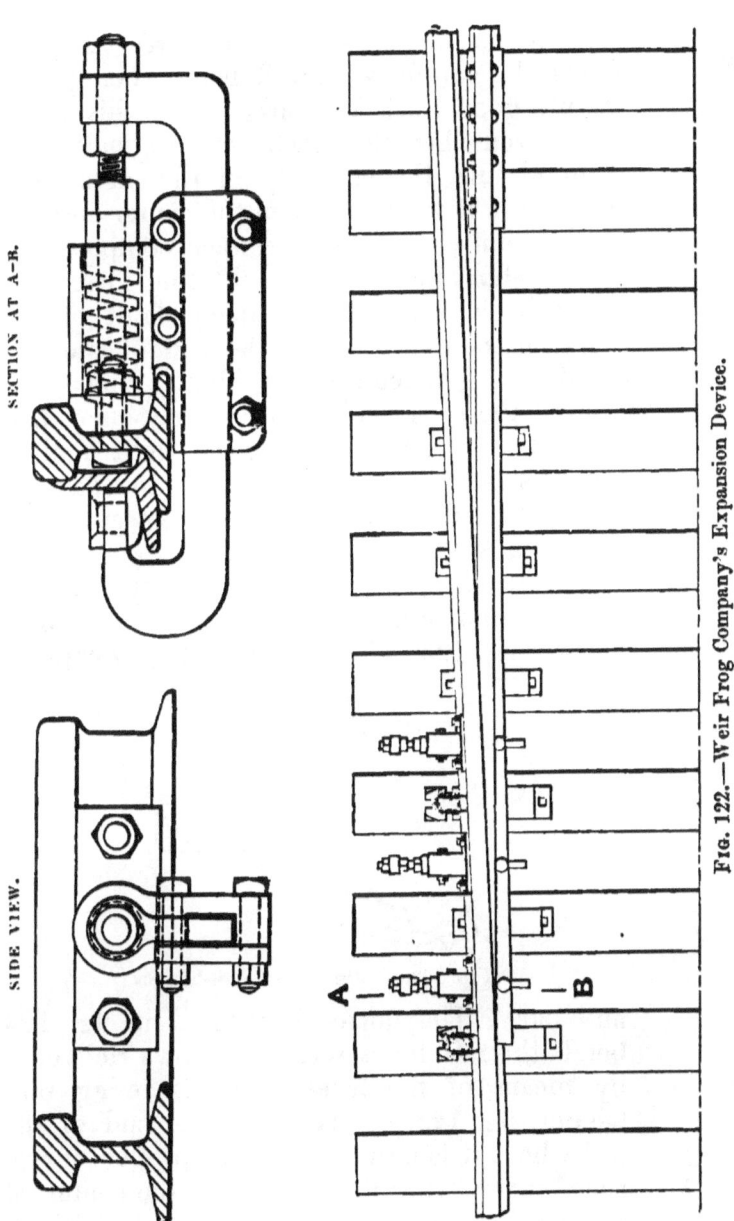

FIG. 122.—Weir Frog Company's Expansion Device.

rail by the springs which may be seen in the sectional view.

Curving rails.
Rails on all curves above three degrees should be carefully curved before being laid when a good track is desired, for if this is not done, even when the track was originally left in good line, the elasticity of the metal will soon cause it to spring in at the center and out at the joints, resulting in a track composed of several short pieces of straight line instead of a regular curve. The frequent fault of not carrying the curve out to the end of the rail should be particularly avoided. And even though the rails have been properly curved, if the spikes have not been driven tight against the rails both inside and outside, the passage of trains will soon develop unexpected and annoying kinks.

Rail benders.
For bending rails there are many devices; the best known and the cheapest is the "Jim Crow," fig. 123, which is worked by a capstan

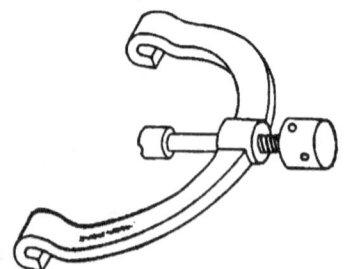

Fig. 123.—"Jim Crow" Rail-bender.

and bar. The device illustrated in fig. 124 bends the rail by a series of blows delivered by means of the lever. The traveling rail-bender, fig. 125, is placed at one end of the rail where it is adjusted to the proper curve, then by revolving the capstan in the center of the mechanism the device is propelled along the rail, curving it as it goes. The hydraulic rail-bender, fig. 126, is an adaptation of the

TRACK WORK. 141

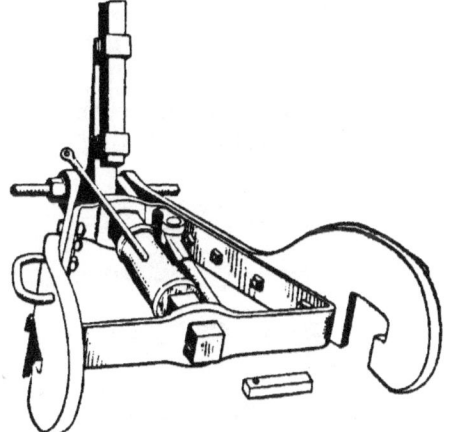

Rail benders.

FIG. 124.—Emerson Rail-bender.
(M. N. Brown.)

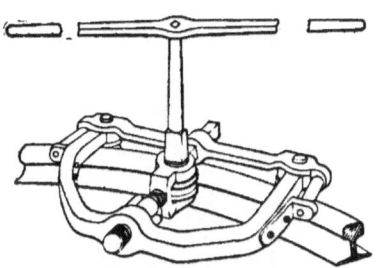

FIG. 125.—Travelling Rail-bender.
(Fairbanks, Morse & Co.)

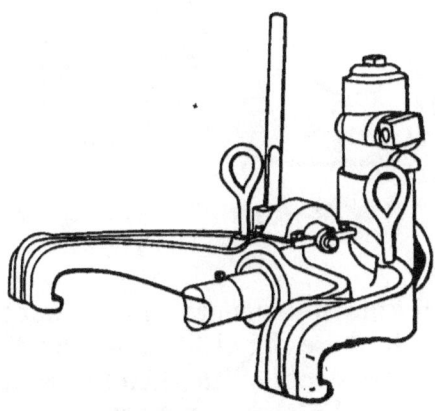

FIG. 126.—Hydraulic Rail-bender.
(Watson & Stillman Company.)

Bolt and spike holes.

hydraulic jack and is an efficient tool, since it operates very rapidly and with great force.
If it is necessary to make holes in the web of a rail to 'be used in the main track it should not be done with a hand punch, fig. 127, except

FIG. 127.—Hand Rail Punch.

under the most powerful necessity; nevertheless the hand punch may be used for side track work and is a handy tool to have on hand when putting in bolts. The hydraulic punch, fig. 128, is a comparatively recent form, while

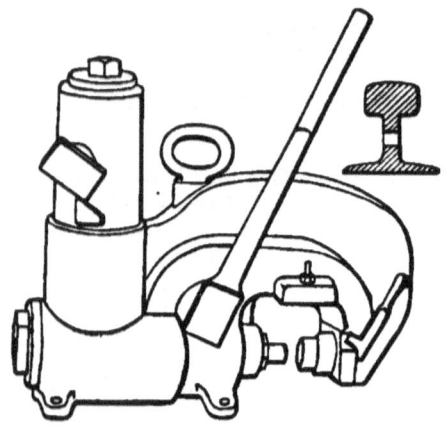

FIG. 128.—Hydraulic Rail Punch.
(Watson & Stillman Company.)

some well known drills are shown in figs. 129 to 134. Of these, fig. 130 is driven forward by both movements of the handle, while figs. 131, 132 and 133 have an automatic feeding arrangement.

TRACK WORK. 143

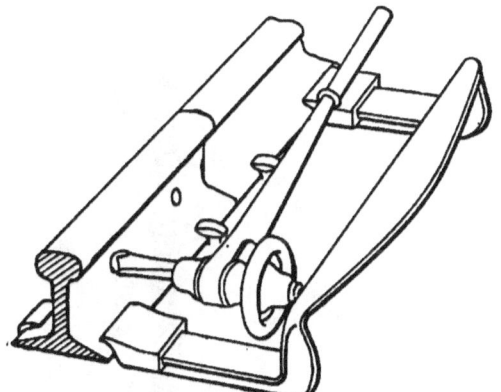

FIG. 129.—Typical Ratchet Drill.

Rail drills

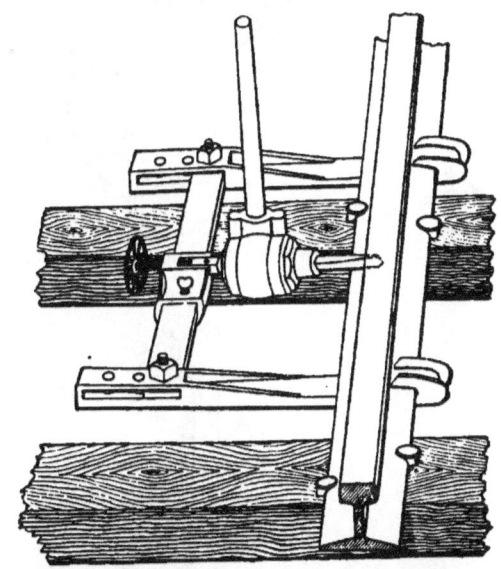

FIG. 130.—Schuttler Double-motion Drill.

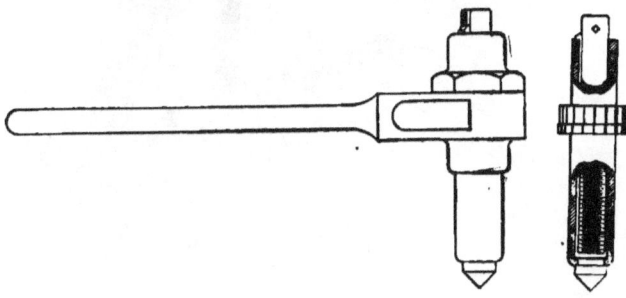

Rail drills.

FIG. 132.—Paulus Drill.
(Buda Foundry & Manufacturing Company.)

FIG. 133.—Buda Drill.

TRACK WORK. 145

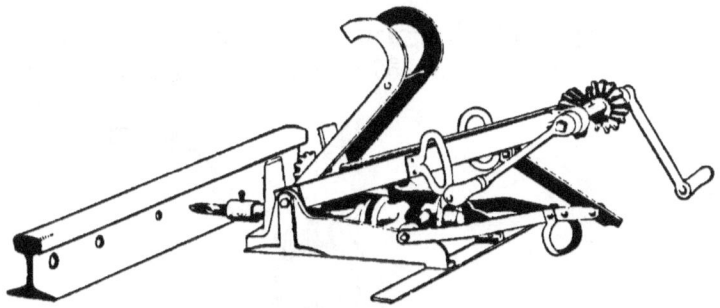

FIG. 134.—The Buda Drill Withdrawn.

An ingenious application of the hydraulic jack is illustrated in fig. 135 for cutting spike

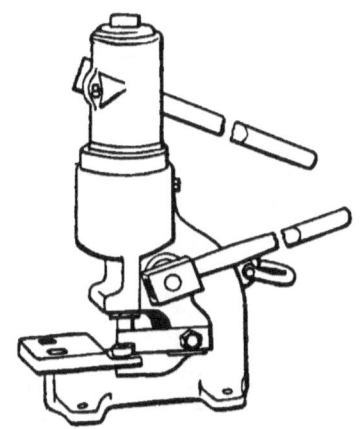

FIG. 135.—Hydraulic Punch for Spike-slot.
(Watson & Stillman Company.)

slots in the base of a rail; this is greatly to be preferred to a chisel or hand punch.

For cutting rails in an emergency and for general rough work the track chisel, fig. 136, **Cutting rails**.

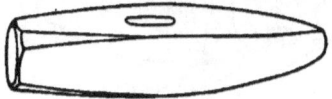

FIG. 136.—Track Chisel.

is likely to hold its own, but like the rail punch it must not be used on the main track. Its best work is coarse compared with that of tools

Cutting rails. especially designed for the purpose as are those shown in figs. 137 and 138. Besides, it is slow,

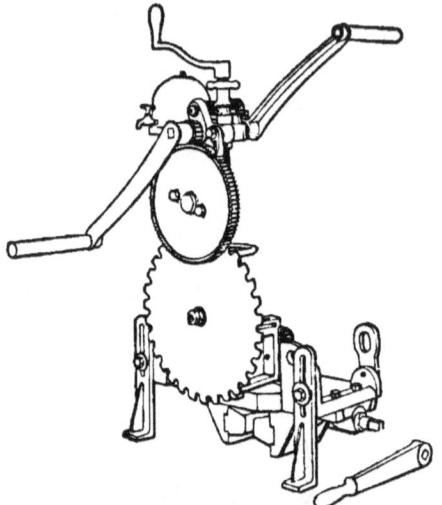

Fig. 137.—Bryant Rail Saw.
(Q. & C. Company.)

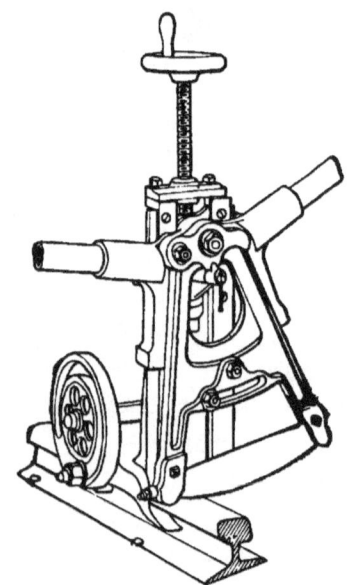

Fig. 138.—Smith Rail Saw.

uncertain and the rail is apt to be bent in dropping it.

CHAPTER XII.

Tools.

There is so large a variety of tools used in maintaining a railroad track that beyond illustrating and describing typical ones of each class, little can be said. Many of them have already been mentioned in previous chapters but many others remain and they will be treated of here.

Each roadmaster, unless the passenger trains are frequent, should have a velocipede-car. Two well known forms are illustrated in figs. 139 and 140, which, however, are likely to be supplanted by the models, figs. 141, 142 and 143. Of these the last two have a distinct

Velocipede cars.

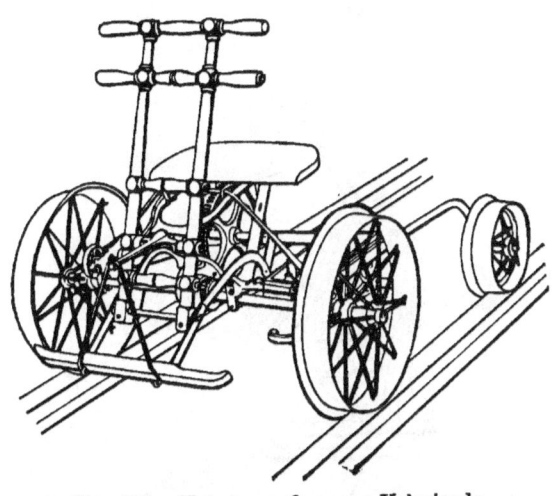

Fig. 139.—Kalamazoo One-man Velocipede.

Velocipede cars.

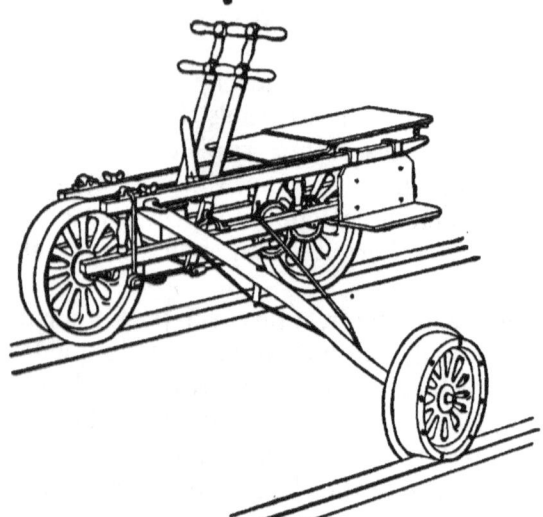

Fig. 140.—Sheffield One-man Velocipede.
(Fairbanks, Morse & Co.)

Fig. 141.—Kalamazoo "Safety" Velocipede.

advantage in that the rider sits in the middle of the machine, from where he sees his track better and is not apt to tip over while going around curves. Convenient and simple locomotive

Velocipede cars.

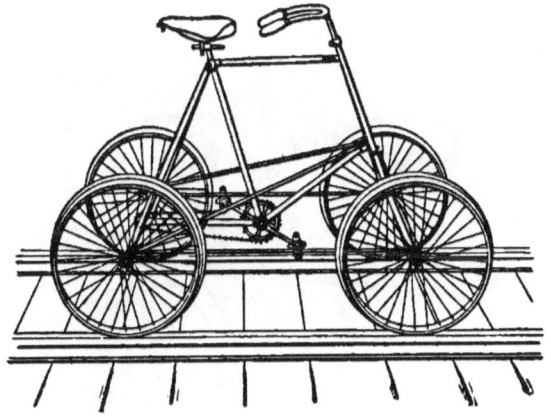

Fig. 142.—Hartley & Teeter Velocipede.

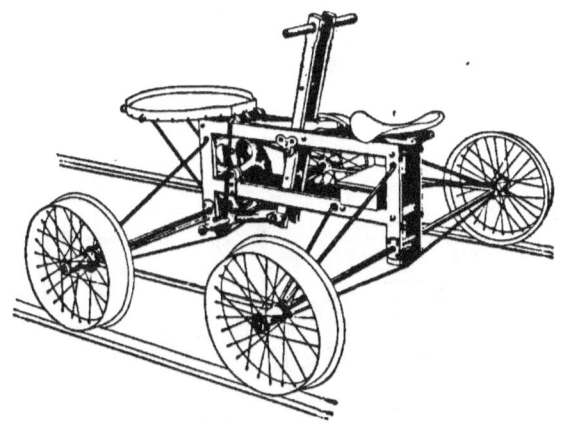

Fig. 143.—Roberts, Throp & Co. Velocipede with Switch-lamp Attachment.

cars are shown in figs. 144 and 145. They are operated by gasoline engines at a cost of (it is said) a few cents a day and attain a considerable speed. They are light in construction and fig. 145 is roomy. It should, therefore, prove of great use for monthly inspections when it is often necessary that two or three persons shall take part. It is not then desirable that the attention of the inspectors should be distracted from their duties by the necessity for performing manual labor, as would

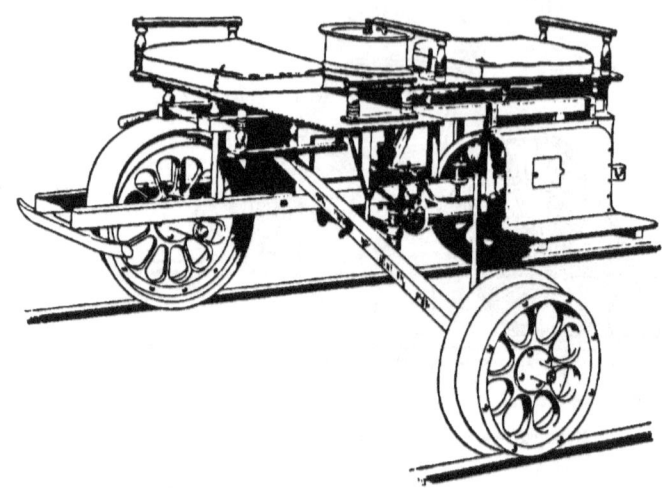

Fig. 144.—Sheffield Gasoline Motor.
(Fairbanks, Morse & Co.)

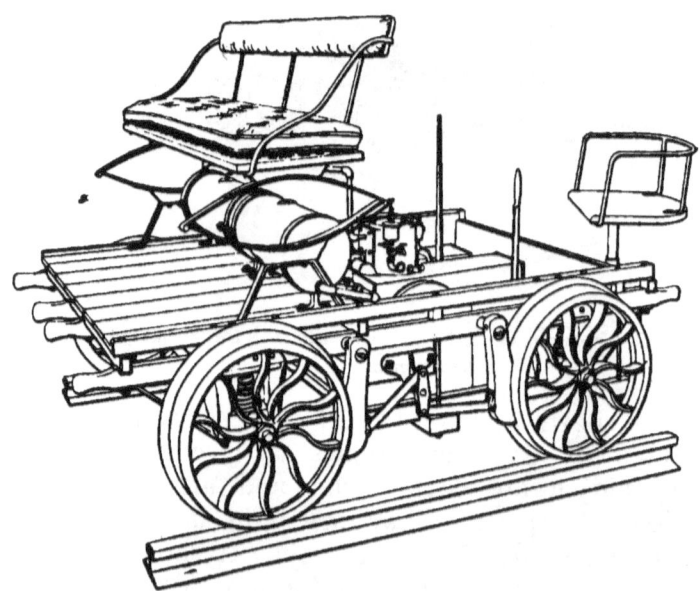

Fig. 145.—Kalamazoo Gasoline Motor.

be the case if a velocipede were used. If the claims made for it are true, it is much cheaper than to take men from a section gang to propel a car.

Each of the sections, except the very short **Hand cars.**
ones, should be provided with an easy-running
hand-car, figs. 146, 147 and 148, as light as is

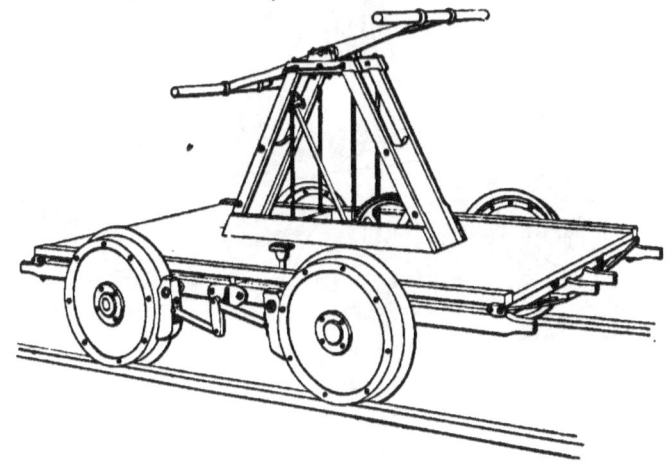

Fig. 146.—Early Kalamazoo Hand Car.

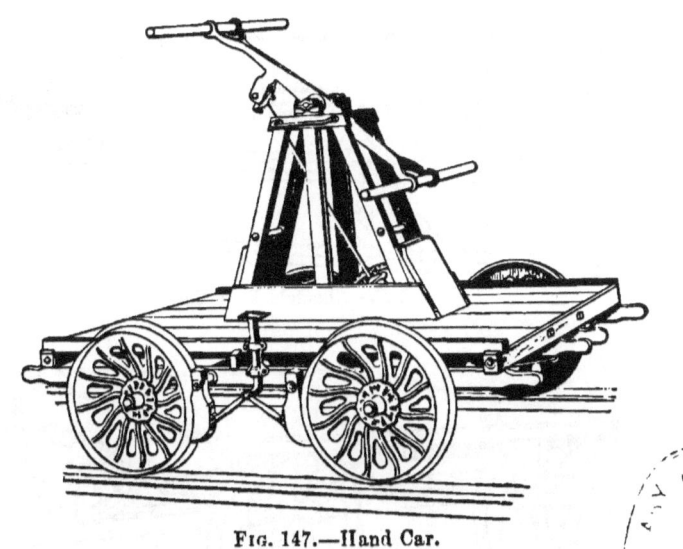

Fig. 147.—Hand Car.
(Buda Foundry & Mfg. Company.)

consistent with strength, for the transportation
of men and tools, *but not material* unless it be
of the lightest. Push-cars of two varieties are
illustrated by figs. 149 and 150. Here strength

152 THE NEW ROADMASTER'S ASSISTANT.

Hand cars.

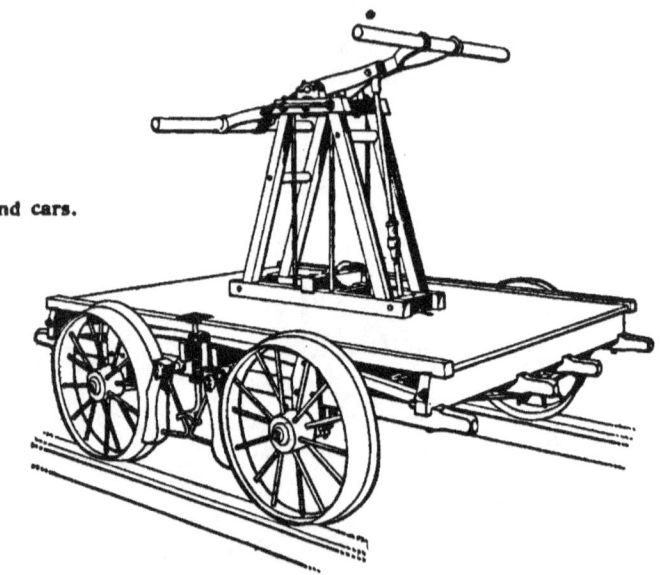

FIG. 148.—Hand Car.
(Roberts, Throp & Company.)

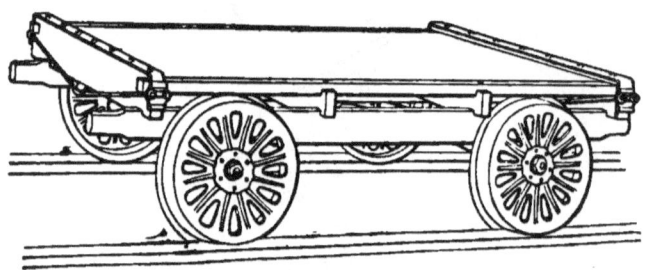

FIG. 149.—Sheffield Push Car.

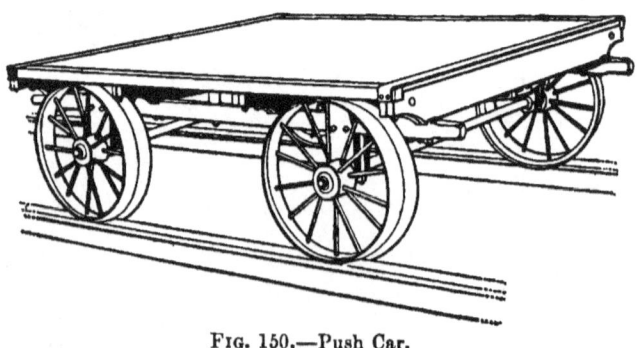

FIG. 150.—Push Car.
(Roberts, Throp & Company.)

and stability are of greater importance than with hand-cars and should be the main feature of their construction.

The track gage should be not only a gage but a square, so formed that, when placed against the rail, it will stand at right angles to it. The old and well known Huntington gage fulfils these conditions. In fig. 151 the lugs

Track gage.

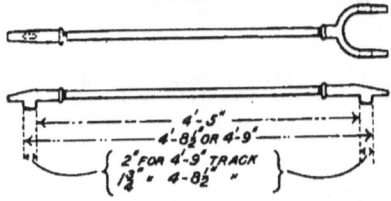

FIG. 151.—Modified Huntington Track-gage.

pointing downward at the ends are made the width necessary to gage the distance of the guard rail from the main rail opposite the point of a frog; this is not a feature of the gage in its original form, but is a modification which will be found convenient and assist in reaching accurate results. All track gages must be compared from time to time with a standard measure kept at headquarters and no track spiking of any kind should be performed without the use of the gage, where the track is unspiked for more than two ties or where the neighboring spikes have been driven for a long time.

The track level, fig. 152, should be substantial but light, formed of white pine, free from

Track level.

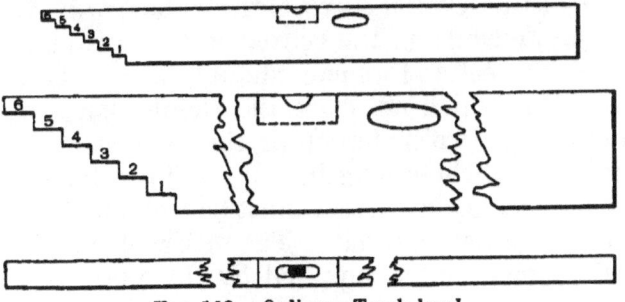

FIG. 152.—Ordinary Track-level.

knots and bound all around with one-eighth-inch iron. Another form is shown in fig. 153;

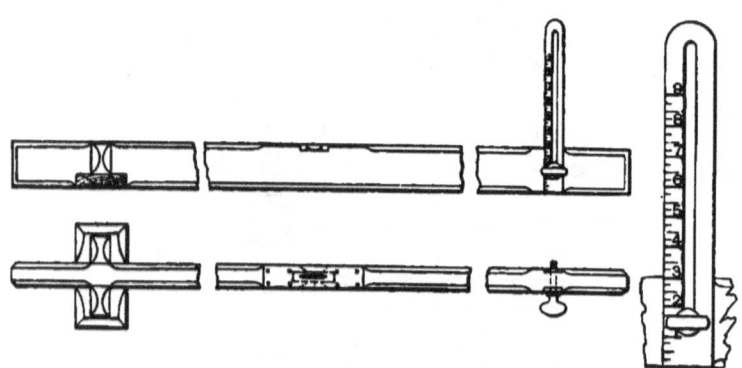

FIG. 153.—Adjustable Track-level.

this one admits of a closer adjustment but is more apt to be damaged and it is likely to prove misleading in the hands of a careless man, or unless it is closely watched. The addition of two gaging-lugs would render it a valuable combination tool for carrying upon the road-master's velocipede car, since it is light and by folding back the slotted scale, becomes very compact. The track-level, like the gage, should be constantly used when working around the track and should be occasionally tested to see that it is really correct. This test can easily be made at any time by setting the level so that the bubble appears exactly in the center of the opening; then after turning it end for end, if the level is correct, the bubble will still remain exactly in the center of the opening.

Tape line. Each section-foreman should have a 50-ft. tape line which need not and should not vary more than two inches from the correct standard in its whole length. These tape lines in order to remain correct must be well made and of substantial material. The miserable printed calico affairs often supplied by the purchasing agent are worse than useless and probably cost nearly as much in the end as a strong and accurate tape. The roadmaster should have in his pocket at all times when on duty, a 25-ft. steel tape. It is an absolutely necessary implement and it will be found useful on many occasions. **Tape line.**

In connection with the tape line, the roadmaster will find the clamp, fig. 154, a useful

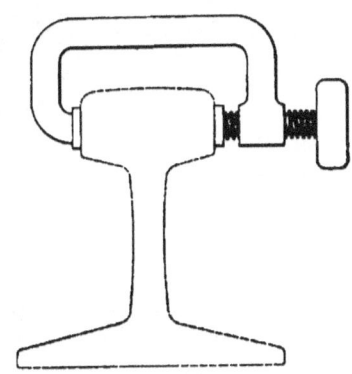

FIG. 154.—Clamp for holding a Tape Line.

and unusually convenient device. It will take the place of a man in making most of the measurements necessary around the track, and with two clamps (which can easily be carried on the velocipede) it is possible to erect and let fall perpendiculars with considerable accuracy.

Of the track bolt wrenches, figs. 155 and 156, it is only necessary to say that, for rapid work, they should be as light as is commensurate with strength and that they should fit easily. It is possible to make the square wrench fit somewhat more loosely than the hexagonal wrench without running the chance of rounding the corners of the nuts, and this is one of the principal reasons for perferring the square nut. The monkey wrench, fig. 157, is subject to such rough usage that it should be made as substantially as possible and because of this the metal handle is better than a wooden one. **Wrenches.**

Wrenches.

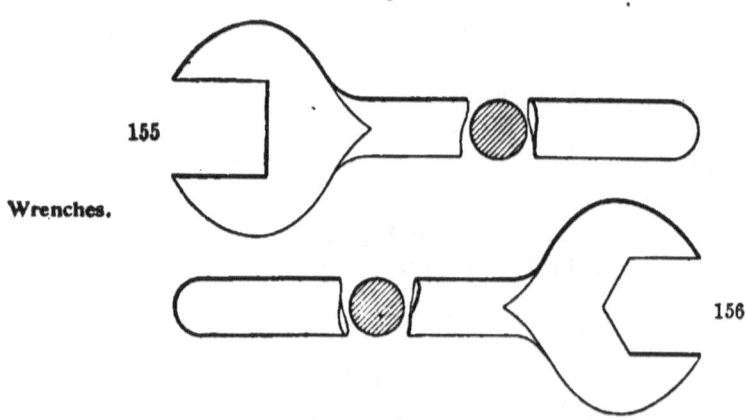

Fig. 155.—Square Nut Wrench.
Fig. 156.—Hexagonal Nut Wrench.

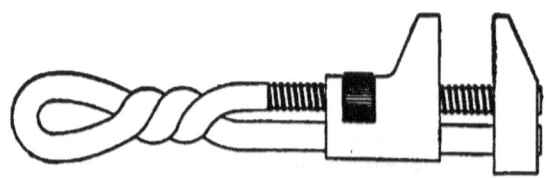

Fig. 157.—All-metal Monkey-wrench.

Mattock. For grubbing and cleaning up a right-of-way and for the rough work to which a finely sharpened axe or adze should not be put, the mattock, fig. 158, is invaluable. Since it

Fig. 158.—Mattock.

combines the useful qualities of two entirely different tools and costs little more than either of them, it presents many obvious advantages.

Picks. The clay pick, fig. 159, need not be heavy but it must be of finely tempered steel, very strong and not much curved.

Picks.

Fig. 159.—Clay Pick.

Fig. 160 represents the best form of tamping tool for stone ballast. It, like the clay

Fig. 160.—Ballast Pick.

pick, must be hard and finely tempered. The weight should be so distributed that the pick will not have a tendency to turn over while the tamping end is in use and this end must be quite heavy and considerably curved in order that the pick may find its way under the tie.

A novel kind of pick is shown in fig. 161 for which many advantages are claimed. It

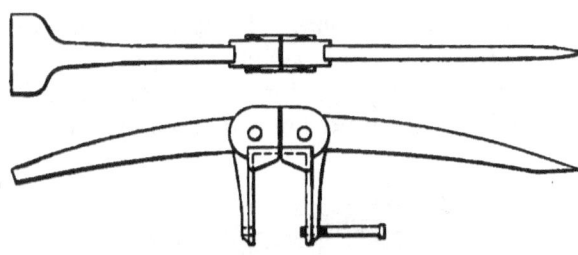

Fig. 161.—Eyeless Pick.
(Eyeless Tool Company.)

may have any form of end as is evident, but the one illustrated is suited only to coarse gravel although the writer has seen it used for stone. It is said that because no eye need be forged in this pick, steel of the same quality may be used throughout and that the pick never needs new points but only sharpening and tempering. The protection afforded by the handle grip is also claimed to increase the life of the handle.

Ballast fork. The only suitable instrument for handling stone ballast is that represented in fig. 162.

Fig. 162.—Ballast Fork.

It is shaped somewhat like a manure fork but is much larger and has square tines.

Tamping bar. Gravel ballast requires different treatment from stone and therefore a different instrument for packing it. The time honored tamping bar is illustrated in fig. 163 and remains the

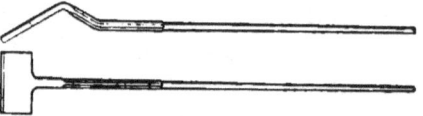

Fig. 163.—Tamping Bar.

same as it has probably been since the first days of railroads and perhaps longer.

Hammers. Napping hammers, fig. 164, for breaking stone ballast by hand should have either very

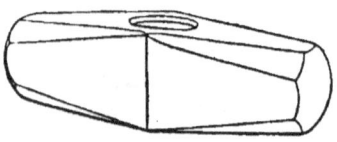

Fig. 164.—Napping Hammer.

long or very short handles depending on whether the men are to stand or be seated while at work. It is usually preferred that they shall stand, but the hammer in either case should be extremely light, not to exceed three pounds without the handle, which should be quite large at the grasp and small where it enters the head. The amount of ballast that a man can make depends very largely on the kind of hammer and handle which he has to work with. Both the napping hammer and a 10-lb. sledge, fig. 165, are a necessity where

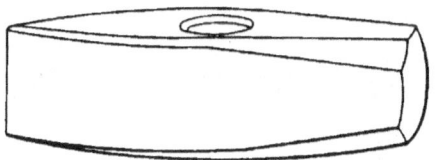

Fig. 165.—Sledge.

stone ballast is used and a convenience on every section.

A spike maul of the ordinary form is exhibited in fig. 166.

Fig. 166.—Spike Maul.

Few devices are better designed for the work which they are to perform, than the spike puller, fig. 167. It will draw a spike

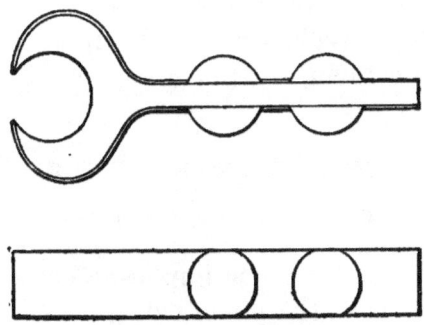

Fig. 167.—Spike Puller.

Spike-drawing tools.

from between a guard rail and its main rail when nothing else will move it.

The claw bar has two distinct and common forms, the "bull's-foot" fig. 168 and the "goose-neck," fig. 169. Most persons who

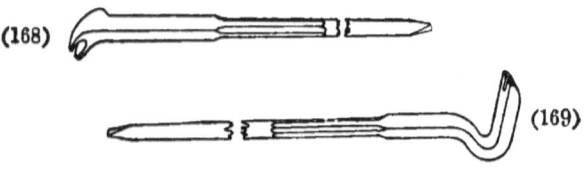

FIG. 168.—Bull's-Foot Claw Bar.
FIG. 169.—Goose-Neck Claw Bar.

have used it prefer the goose-neck, since it has a longer reach and does not require a spike or a stone to be placed under it when a spike is half pulled. Neither is the goose-neck so apt to bend the spikes in drawing them as is the bull's-foot.

Lining and raising bars.

The lining bar, fig. 170 is particularly intended for lining and surfacing track and it

FIG. 170.—Lining Bar.

should therefore be quite long and heavy. It is not suited for much of the work which is required of a crow-bar and if crossing planks or platform planks are to be raised without destroying them, the pinch bar, fig. 171, must

FIG. 171.—Pinch or Raising Bar.

be used. As its name implies, a car can also be moved with the pinch bar.

Shovels.

The snow shovel shown in fig. 172 is of simple construction, strong, durable and is

TOOLS. 161

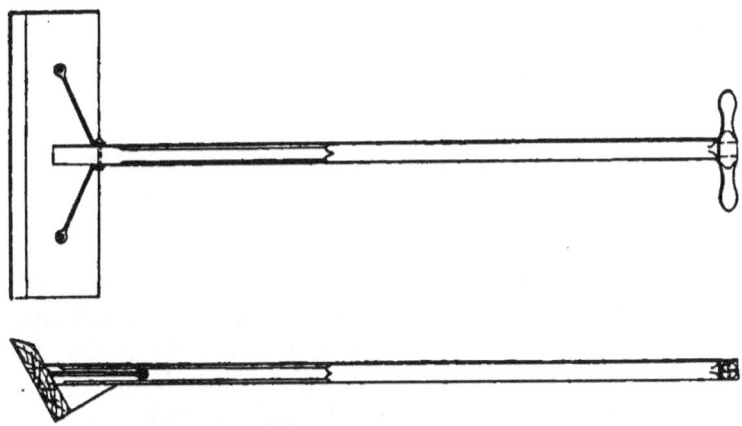

FIG. 172.—Snow Shovel.

particulary intended for cleaning platforms or bringing snow into piles whence it may be shovelled into a car and taken away. **Shovels.**

The dirt shovel, fig. 173, is for the ordinary work of a section.

FIG. 173.—Dirt Shovel.

The long-handled, sharp-pointed ditching shovel, fig. 174, is the one which will be found

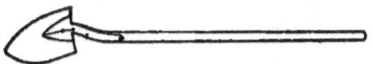

FIG. 174.—Ditching Shovel.

most generally useful for its purpose. It will take up as much dirt as a man can properly throw from a ditch to a partly loaded flat car; it penetrates the earth easily by reason of the point and it will take a sufficiently thin slice. Fig. 173 is also made with a long handle but, except for some particular reason, is not usually required on a section.

Shovels. Fig. 175 shows the long, narrow shovel best adapted to tile drain work; because of

FIG. 175.—Tile Drain Shovel.

its shape, a trench scarcely wider than the tile drain, may be excavated to a considerable depth, with a corresponding saving in the quantity of material which must be moved.

For digging post holes, fig. 176 represents the form of tool most generally useful. In

Post holes.

FIG. 176.—Post Hole Shovel.

soft, even soil however, the "scissors" digger, fig. 177 works very rapidly. This is driven

FIG. 177.—Post Hole Digger.

into the ground with the handles as they are shown in the drawing; they are then spread apart (which brings the scoops together), the digger is withdrawn and the earth emptied onto the ground.

Rail tongs and fork. It seems unnecessary to do more than mention the rail tongs, fig. 178, for carrying rails

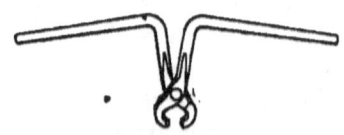

FIG. 178.—Rail Tongs.

TOOLS. 163

and the rail fork fig. 179 for turning rails on cars or when it is desired to inspect them.

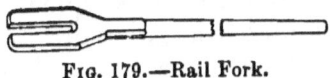

FIG. 179.—Rail Fork.

A flag and lantern-holder of some sort is **Flag-**
necessary in raising track and a simple arrange- **holder.**
ment is shown in fig. 180. It is formed of a

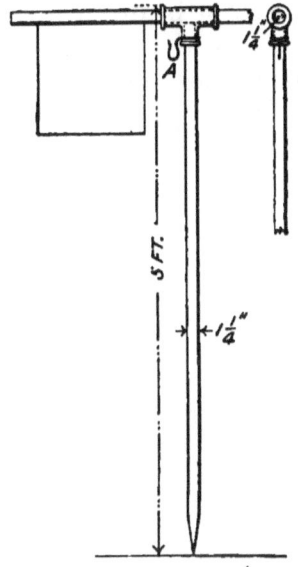

FIG. 180.—Flag Holder.

piece of gas-pipe about six feet long, pointed at the bottom and with an ordinary cast-iron "tee" screwed to the top. The lantern is supported on the hook at A which is bent around the "tee" and is made of $\frac{1}{4}$-in. round iron. If a hole be bored through the end of the flag stick, and a pin be put through the hole after the flag has been placed in the holder, the flag will be prevented from falling out while a piece of telegraph wire, bent into the proper shape, will, if sewed into the edges of the cloth, prevent the flag from being rolled up by the wind.

Care of tools.
A number of tools should be assigned to each section sufficient to keep all of the men employed at any kind of work that may be going on and to replace dull tools which have been sent to the blacksmith shop for sharpening. Before being issued the tools must be branded or stamped with the initials of the railroad company in such a way as to make it impossible to efface them without destroying the tool, and except in rare cases, a track-foreman should be forced to turn in his old tools at the time new ones are issued to him. The tool report well repays attention, but it is the one most frequently lost sight of.

CHAPTER XIII.

Frogs, Switches and Switch-Stands.

In the chapter following this, and among other rules and tables, will be found a simple, diagrammatic method of laying out switches, together with explanations of frog numbers, angles, etc.; here we shall deal only with the physical characteristics of various classes of material.

It is necessary first, however, to state that all switch nomenclature is based directly upon what is known as the "number" of its frog. In other words, a number 6 switch lead is that lead which would be used with a number 6 frog. *Frog angles and numbers.*

There are many kinds of switches (almost all of them patented), which it is impossible to describe here. Most of them are intended to keep on the track a train which is passing over an open switch in a trailing direction. From the fact that none of them, except one (the Wharton) has been widely used, it would appear that they have not proved more valuable than the simple split switch which is now in almost universal use on main tracks in the United States. *Varieties of switches.*

The Wharton switch was designed to accomplish two things in particular: first, to provide an unbroken rail for trains on the main track; second, for protecting trains approaching an open switch in a trailing direction, and these were quite successfully done but at a cost which caused its abandonment in most localities. One form of this switch, the "Robinson-Wharton," illustrated in fig. 181, differs *Wharton switch.*

166 THE NEW ROADMASTER'S ASSISTANT.

Robinson-Wharton switch.

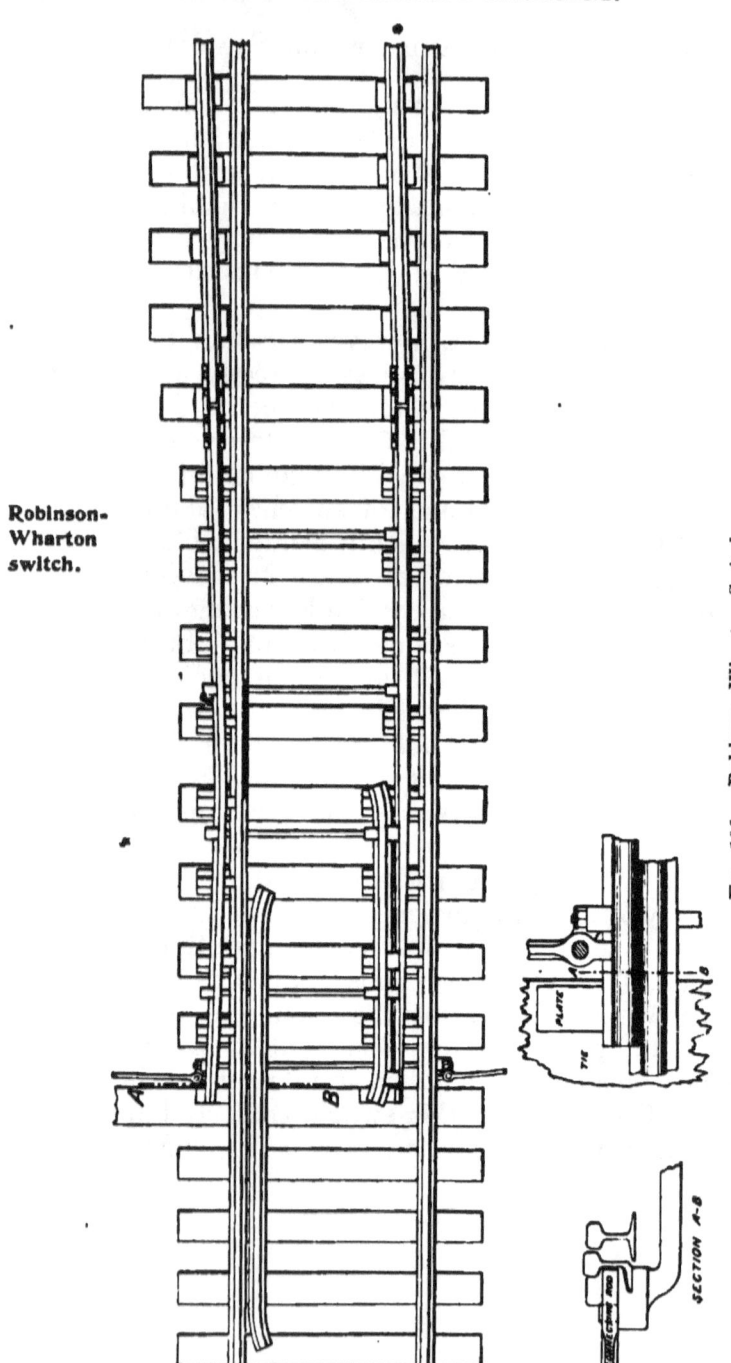

FIG. 151.—Robinson-Wharton Switch.

considerably from the original, in having its two moving rails made from ordinary T-rails instead of from specially shaped materials.

Split switches. Improved designs and processes of manufacture, added to the general use of steel in railroad tracks are the causes which have tended to produce so simple and satisfactory an article as is a well-made split switch. There is illustrated in fig. 182 A and B a typical split switch, combining in many of its details the best methods known at the present time. Some trackmen prefer flat rods, and some round rods; some wish to have the jaw of the rod placed on the switch rod fastenings and some prefer to have it on the switch rod itself. Generally speaking, the less moveable parts there are about a switch the better, and in the design above mentioned there is not a single bolt, nut or weld. The fastenings are riveted to the rail, the rods are formed of two pieces of flat iron riveted together, which, being separated at the end, form jaws. The rods and feet are connected by turned pins furnished with cotters, while the connecting rod and switch rod are also joined by a turned pin and cotter.

Switch rods. The common practice of using four and even five rods in a split switch, is an inheritance from the old stub switch, which depended on the rods to hold the rails together. In a split switch however, the rail which is in use, can and should be always braced on the outside for more than half its length; this renders more than one rod an unnecessary complication. The only purpose that the additional ones could possibly serve would be to hold the rails together in case of a break and it is hard to see how they could do even this much. In any event there is a much better plan, that of reenforcement, and one

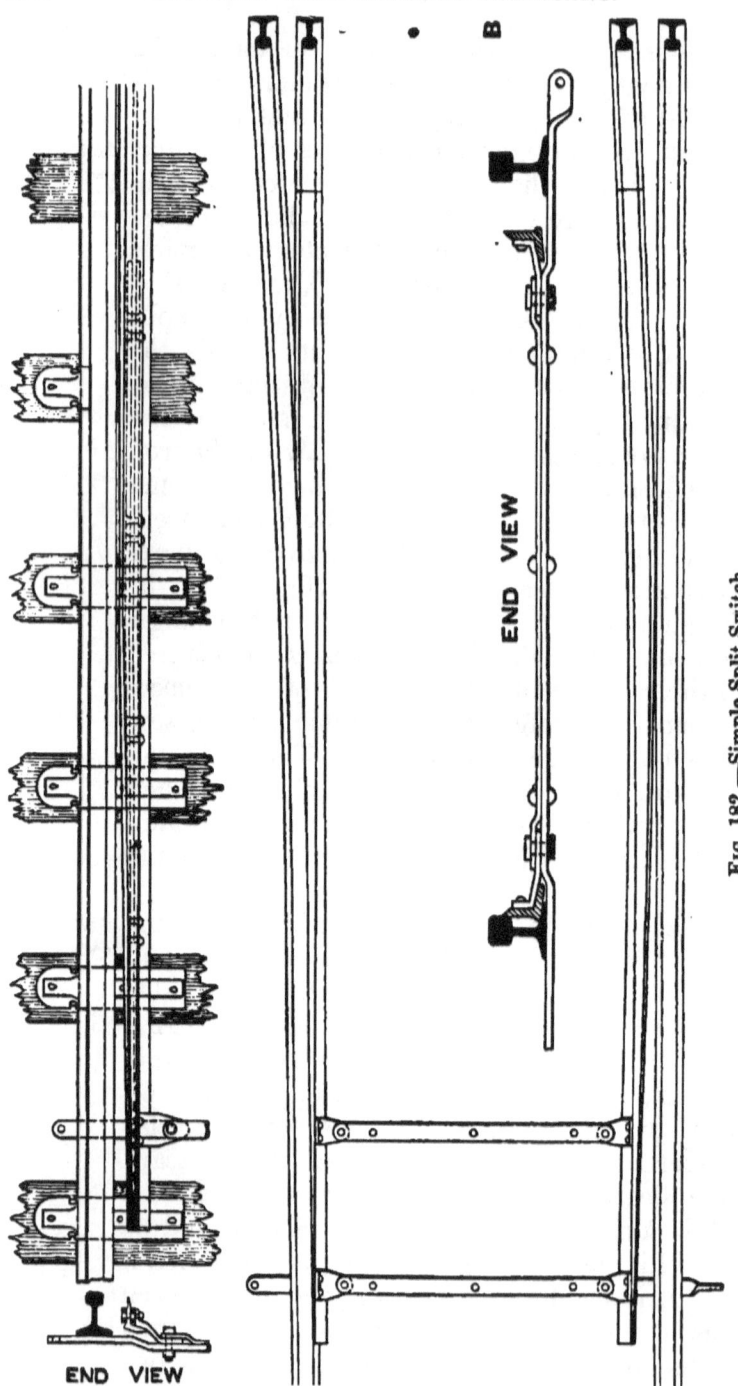

Fig. 182.—Simple Split Switch.

wonders why any of the old four or five rod switches are built.

Two ways of reenforcing split points are shown. The first, at C in fig. 182 is the most obvious method and is accomplished by rivetting through and on each side of the web, two bars of steel. Although this is probably inferior to the plan shown in fig. 183, it is far stronger and more reliable than the old method. The plan illustrated by fig. 183, has a plain bar on the side next to the main rail but the bar on the other side is replaced by a strip of angle steel to which the switch rod is attached. It is evident that the horizontal flange of the angle is much stiffer laterally and lighter than the flat form of metal. Two particular advantages which are to be expected from the reenforcement of split points, through the removal of the switch rods, are the lessening of damage from dragging brake-beams and the absence of interference from snow and ice. *Reenforcement.*

Fig. 184, although not a split switch in the true sense of the term, resembles it enough to entitle it to a place in the same class. It is rather heavy, since but little of the head and apparently none of the base is removed; at the same time its construction leads one to expect great strength and durability. *Stewart switch.*

Three-throw switches have always been the bane of a trackman's life. The yardmen like them because they save running about and because that other blessed nuisance, "a flying switch" can be made through them so easily. They are always to be avoided if possible and usually can be, particularly in main track, although sometimes a set of circumstances will be met where no other solution of the problem is evident. By means of the reenforcement of the points, this form of switch has been robbed *Three-throw split switches.*

of at least a part of its terrors and a three-throw switch of this sort is shown in fig. 185.

Three-throw split switches.

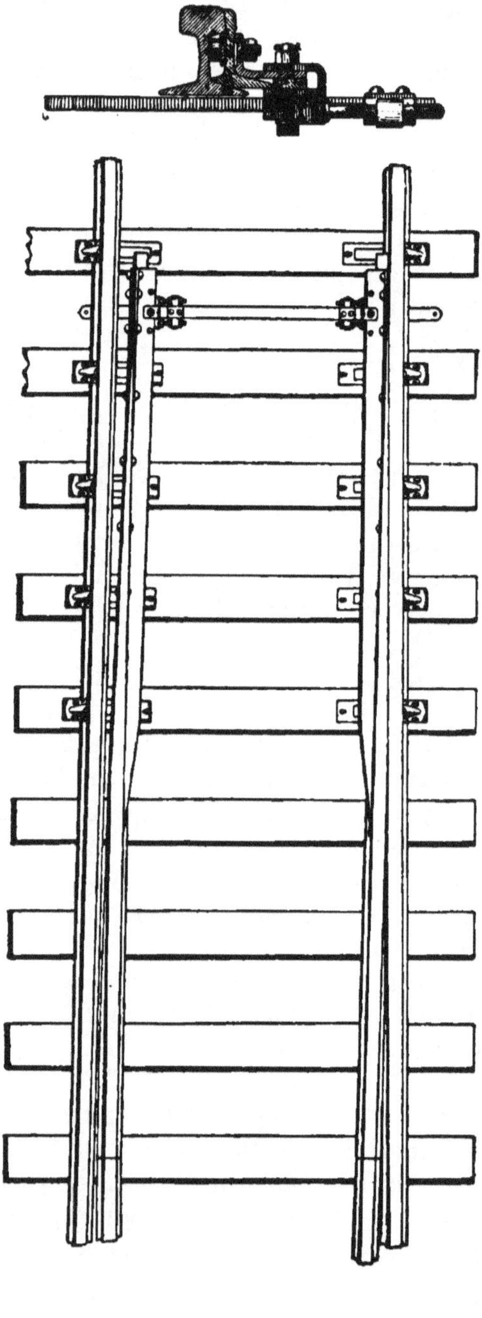

Fig. 183.—Pennsylvania Steel Co.'s Reenforced Switch.

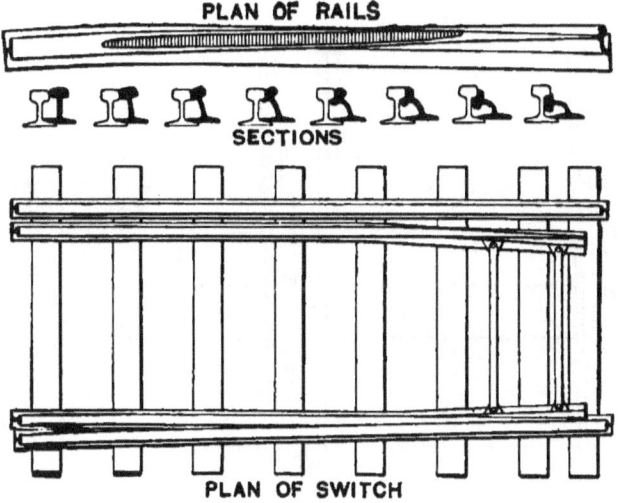

Fig. 184.—Stewart's Switch.

Throw of switches.

The throw of split switches should not be less than 3½ inches nor more than 5 inches. On some railroads five inches is preferred, for the reason that all switch stands, whether for stub or split switches may then be interchangeable but the increasing use of split switches even in side tracks and the fact that for all interlocking the throw must be small leads to the belief that 3½ inches is the better distance.

Adjustment.

The adjustment of a switch may take place by either of two methods. The first plan is presented in fig. 186, where there is a plate (shown in the illustration) containing two holes by means of which it is riveted to the switch rod. The ends of the plate are also bored with large holes and bent in at right angles. Through them are loosely fitted two sleeves which in turn are mounted upon the connecting rod and are held in position there by four nuts, one on each end of each sleeve. The switch rod may move through the plate only an amount depending on the distance

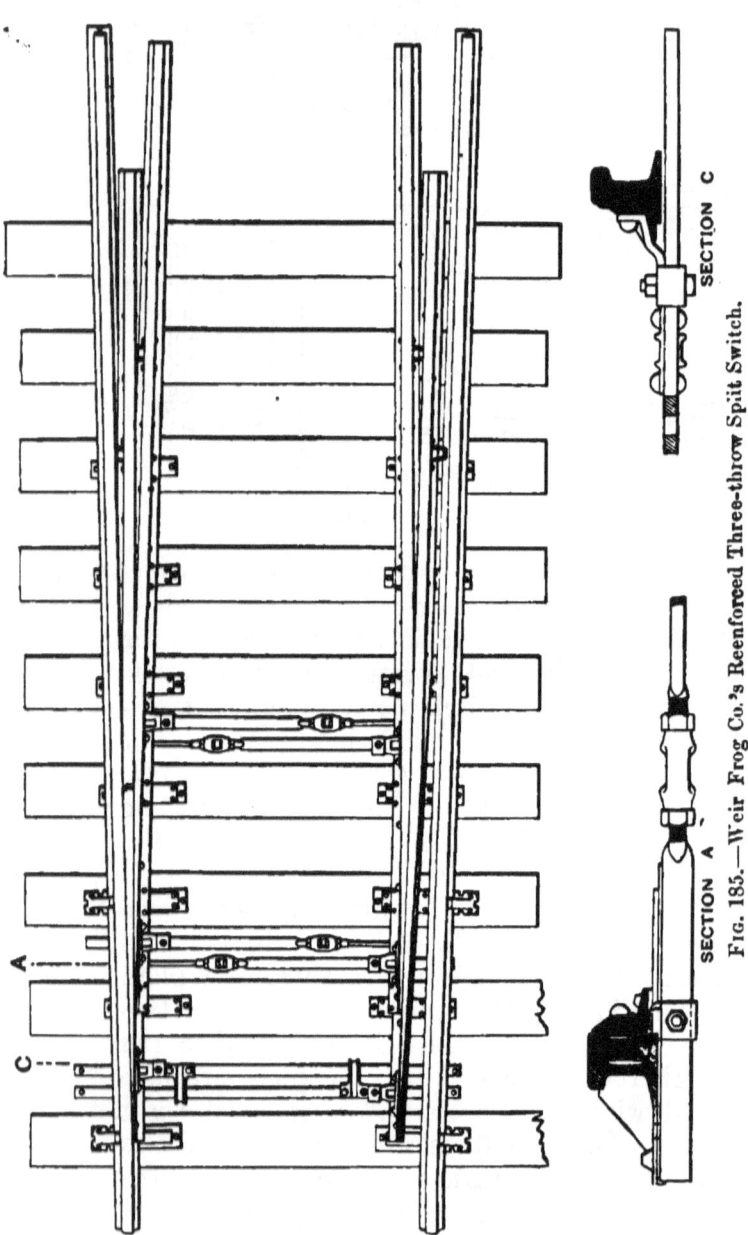

Fig. 185.—Weir Frog Co.'s Reenforced Three-throw Spit Switch.

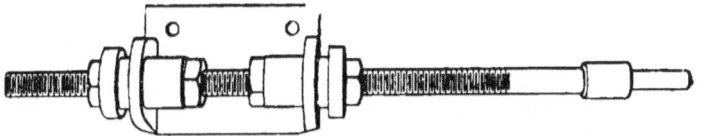

Fig. 186.—Union Switch and Signal Co.'s Switch-throw Adjustment.

apart of the sleeves, since it is stopped on both sides by offsets on the outside end of each sleeve. The switch stand must always have a throw somewhat greater than the throw of the switch and one advantage of this method is that it may be considerably greater. If an adjustment is necessary the two sleeves are moved away from or toward each other depending on whether the throw of the switch is to be diminished or increased. A distinct disadvantage of this plan is the ease with which the adjustment can be made (it is only necessary to turn the nuts) whereas, next to destroying the switch, the adjustment should be made as inconvenient as possible, since in that way the chance of someone's tampering with it is reduced. But on the other hand there is with this plan an unbroken switch rod, a condition not possible in the method next to be described. Fig. 187, presents the second

Switch-throw adjustment.

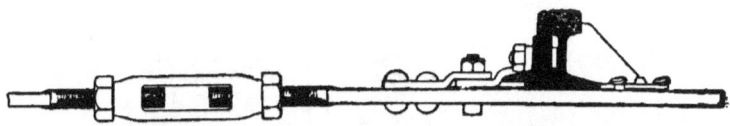

Fig. 187.—Weir Frog Co.'s Switch-throw Adjustment.

plan and that is by means of a turn-buckle, but it must not be one of the ordinary form which has a right hand thread in one end and a left hand thread in the other. This must have either a right or left hand thread at both ends; then if the jam nuts become loose, the turn-buckle will revolve perhaps, but it cannot change the throw of the switch. It is therefore evident that in order to adjust by this method, one end of the rod must be removed from its rail.

Rail braces. Not less than five rail braces as in fig. 182, should be used on each side of a split switch and in connection with them tie plates should always be provided to prevent the rail from cutting into the tie and to form a surface on which the split rail may move easily.

Point guard-rails. If the track is properly gaged and the switch properly put in, guard rails at the point should not be necessary. Since the throw of a split switch should never be less than $3\frac{1}{2}$ inches, a wheel which would catch the open point would scarcely pass over the numerous highway crossings which exist, without being derailed at one of them.

Stub switches. The stub switch is composed of ordinary T-rails with two head chairs and some switch rods. It is almost unnecessary to say that it should never be used for main track as it is dangerous in many ways and is extremely difficult to maintain under a heavy traffic. The best practice requires that the head chairs, fig. 188, shall be of wrought iron, wrought

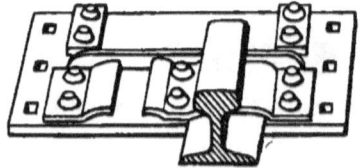

FIG. 188.—Wrought Stub Switch Chair.

steel (not cast) or malleable iron; that two bridle rods shall be used and that the switch rods shall be formed of not less than $1\frac{1}{2}$-in.

round or square metal. The throw should be 5 in.

The arrangement of main track and side track guard rails at frogs is shown in fig. 189.

Frog guard rails.

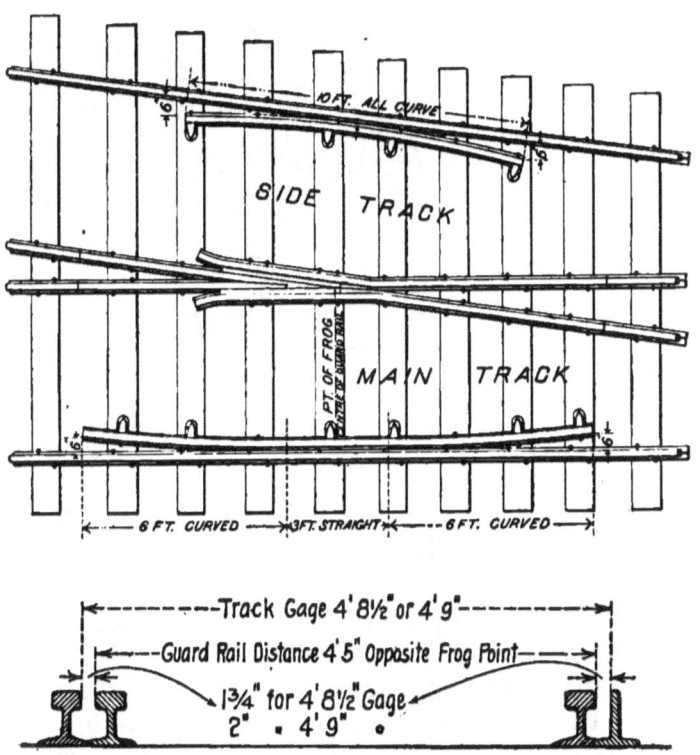

FIG. 189.—Arrangement of Guard Rails.

They should be respectively 15 ft. and 10 ft. long, with their centers opposite the point of the frog. Three feet of the main track guard rail at the middle portion should be straight, spaced 2 in. from the main track rail for 4 ft. 9 in. track, and $1\frac{3}{4}$ in. for 4 ft. $8\frac{1}{2}$ in. track. Side track guard rails should be curved in the same general way, but the straight piece three feet long should be omitted. Not less than six braces should be used on

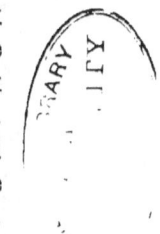

Guard rails.

main track nor less than four on side track guard rails. As a substitute for the rail brace (figs. 119 and 120) at guard rails, the clamp, fig. 190, is suggested. Although more ex-

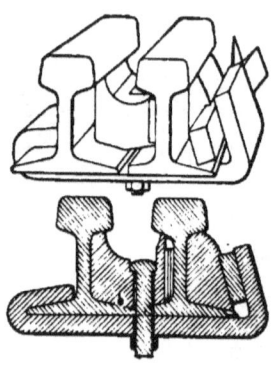

FIG. 190.—" Standard " Guard Rail Fastener.

pensive, it is much more certain and efficient than the rail brace.

For spacing guard rails as well as for all other track spiking, a gage is necessary, with lugs having a width equal to the distance between the guard rail and the main track rail heads. Such a gage, represented in fig. 151, is simple in form, inexpensive and more important still, because of the yoke, it acts as a square.

Foot guards.

Many states require that employees shall be protected from the danger due to frog openings, but whether the law on the subject is operative or not, there is a moral obligation which railroads cannot afford to ignore.

These openings are sometimes closed with wedges of wood which are a makeshift at best and there are other arrangements which are built into the frog and may be purchased of any switch and frog maker. Still other forms

FROGS, SWITCHES AND SWITCH-STANDS. 177

which are removable are illustrated in figs. 191 **Foot**
and 192. **guards.**

End view.

Side view.
FIG. 191.—Foot Guard for Frogs and Switches.
(Roberts, Throp & Co.)

Of these, fig. 191 is provided with springs for the purpose of holding it tightly in position. It is made in several different forms to suit the particular place which it is to fill.

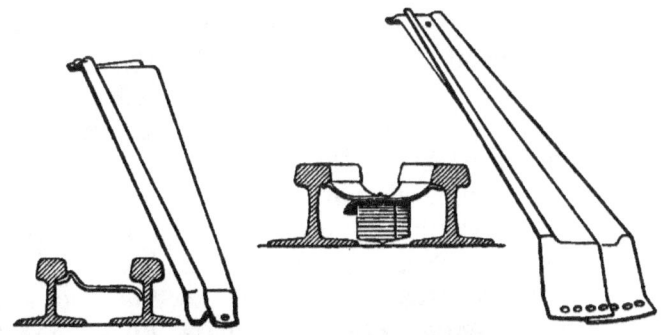

FIG. 192.—Sheffield Foot Guard.
(Fairbanks, Morse & Co.)

That detail shown in the left-hand portion of fig. 192, is the form built for the wing spaces of frogs and the heels of switches. The right-hand detail is for the crotches of frogs and is built of two pieces, hinged at one end in order that any guard may be adapted to a crotch of any number.

Slip switches. The "slip switch" is illustrated in two forms in figs. 193, A and 193, B. Its main object

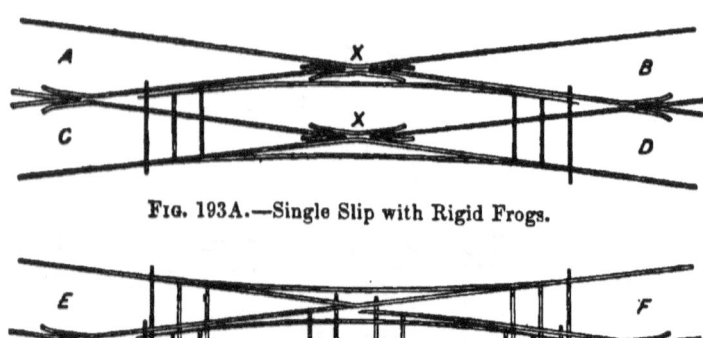

Fig. 193A.—Single Slip with Rigid Frogs.

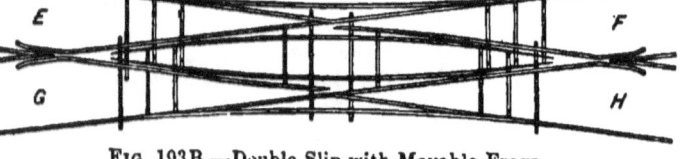

Fig. 193B.—Double Slip with Movable Frogs.

is for the economizing of track room, and this it does in a complete and beautiful manner. Fig. 193A, is what is known as a "single slip with rigid frog" and evidently provides two routes from both C and D, but only one route each from A and B. In fig. 193B, which is called a "double slip with movable frog" two routes are possible from all four of the entrances E, F, G, H.

For general work the best length is that which would be used in connection with a No. 7 or No. 8 frog. If a frog of larger angle is used, the curves become too sharp for passenger train movements, unless the gage of the track is widened. When the frog angle is smaller than No. 8, it becomes necessary to use movable (fig. 193B) frogs, since trains are apt to be derailed by taking the wrong side of one of the rigid double pointed frogs, X in figs. 193, A and 195. The movable frog is also a simpler, cheaper, better and safer device than the rigid frog.

There are few places where either of figs. 194 and 195 may be used (for they are inter- **Movable frogs.**

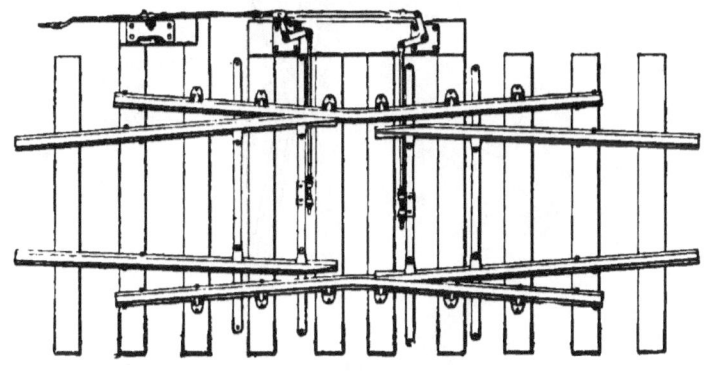

FIG. 194.—Movable Frog.

changeable in that both the movable frog and the rigid frog, ✕ in fig. 195, are used in similar places) at which the movable frog would not be the better of the two. The movable frog requires no guard rails because there is no opening at the point, whereas the rigid frog has and must always have openings between the points, and these cannot be protected. The movable frog is nothing but two bent rails and two sets of planed points; it may be used in connection with either of figs. 193, A and 193, B or alone, as at a simple crossing; last but not least, it furnishes a continuous rail and makes the track as smooth as at a split switch.

For larger angles than are possible with the movable frog some other form is required if the track is to be continuous. Many attempts have been made to fulfill this condition but without great success, commercially at least. But if grade crossings are to prevail in this country some means is needed to improve their construction and behavior. The most rational **Continuous rail crossings.**

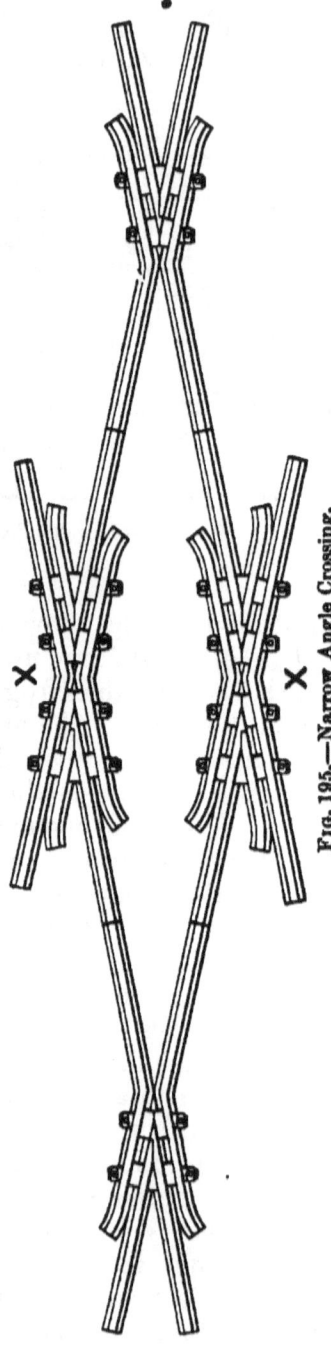

Fig. 195.—Narrow Angle Crossing.

FROGS, SWITCHES AND SWITCH-STANDS. 181

and substantial of the continuous-rail, wide-angle crossings is represented in fig. 196, and **Continuous rail crossings**

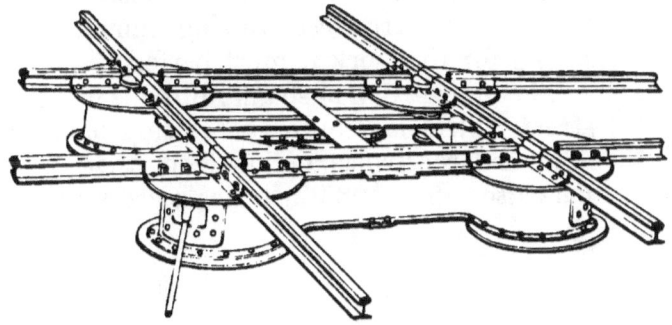

FIG. 196.—Fontaine Crossing.

consists of four revolving turrets, supported in a wrought-steel frame and carrying rail heads on their tops.

The narrow-angle crossing has already been shown and spoken of in connection with fig. 195. The wide-angle crossing is illustrated in fig. 197. It is impossible here, to discuss the **Rigid crossings.**

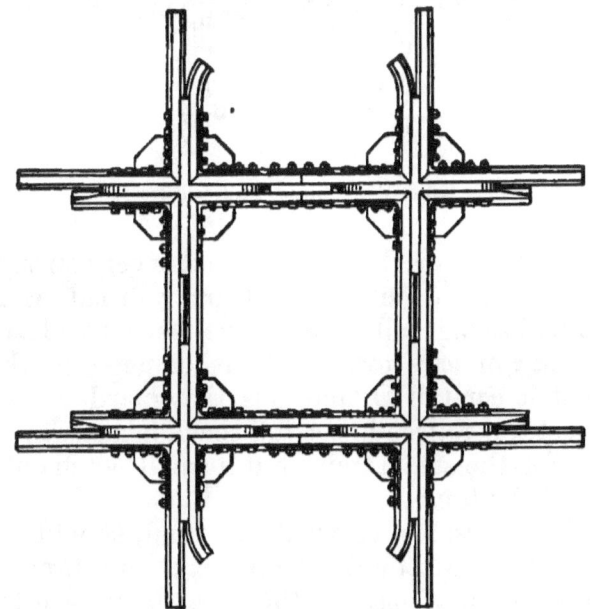

FIG. 197.—Wide Angle Crossing.

different methods of construction, which are numberless, complicated and would require a book in themselves. Suffice it then to say that the heaviest, strongest crossing cannot be too strong for the work it must perform.

Street railroad crossings. The intersection of steam railroads by electric street railroads requires a crossing of a somewhat different construction from that in ordinary use. Fig. 198 illustrates a crossing which

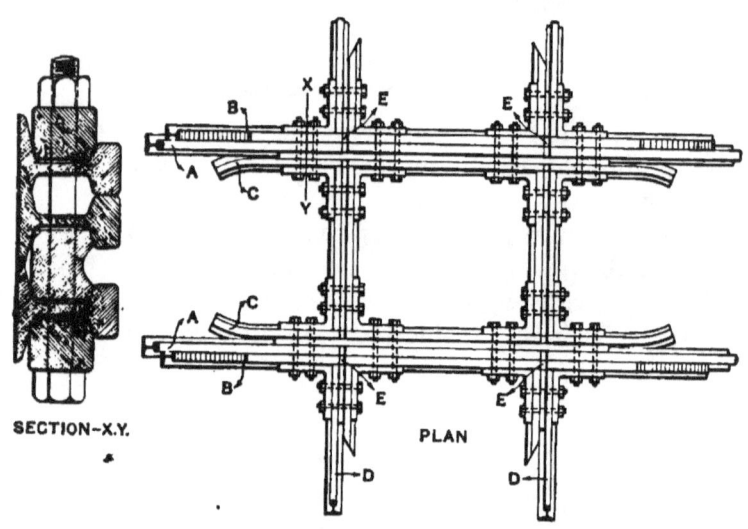

FIG. 198.—Steam Railroad and Street Railroad Crossing.

is designed for this purpose and is very strong. A is the main rail of the steam railroad, B is a reenforcing rail to carry the worn treads of the steam cars over without damage to the electric car rail D, and C is the guard rail of the steam railroad. A full-sized flange-way is left for the steam trains but a small notch only is cut for the electric cars at E.

Crossing foundations. All crossings, of whatever kind, should be placed on substantial white oak and broken stone foundations, for they are the most difficult parts of a railroad track to keep up.

The single pointed or "switch-frog" is made **Switch** in two general forms, commonly known as **frogs.** "rigid" and "spring-rail" and these are again subdivided according to the way in which the rails are put together. Figs. 199, 200 and 201

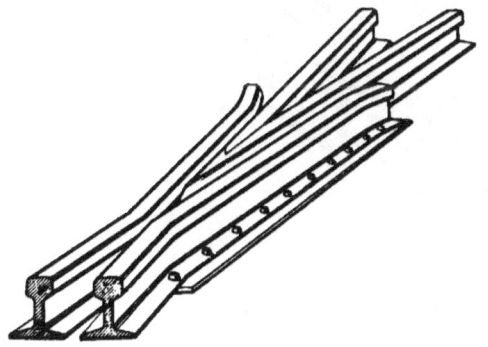

FIG. 199.—Rigid Plate Frog.

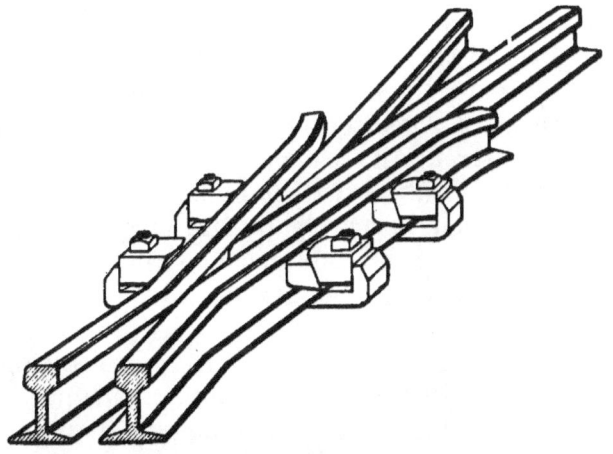

FIG. 200.—Rigid Yoke Frog.

show respectively a "riveted plate" frog, a "clamped" (or "yoke") frog and a "bolted" frog, all of them "rigid." Of these three forms the "yoke" is probably the best and the "plate" frog the least desirable, since the ties must be cut out to receive it.

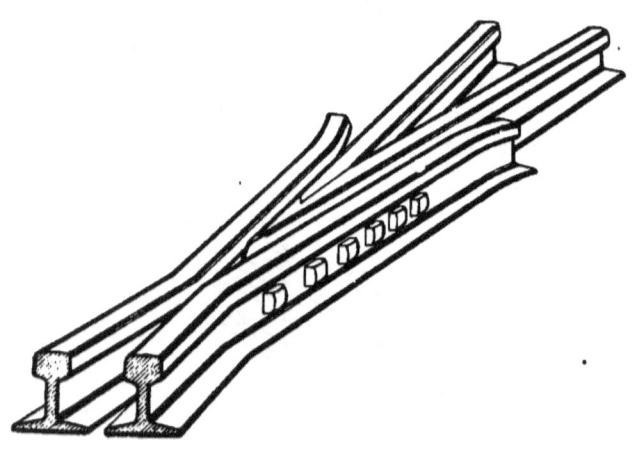

Fig. 201.—Rigid Bolted Frog.

Spring-rail frogs. The spring rail frog, which is intended to furnish a continuous rail on the main track, is also built as a "plate," "yoke" and bolted frog. Through faults of design, as is believed by many persons, the spring-rail frog was the cause of some serious wrecks, which did much to discredit its use on several important railroads. But notwithstanding this it has been continued in service by the larger portion of our lines, until now its construction has been so improved, and its parts so strengthened that it is safe to recommend its use when properly designed and built. A typical spring rail frog is illustrated in fig. 202. This particular form (which is the commonest) requires that the spring rail T H shall be unspiked and free to move sideways. held only by the splices at T, the springs S S and the guides G G. To obviate this necessity another method has been devised, illustrated in fig. 203, which permits the main-track rail to be fastened to the ties as far as 'P. The spring rail is pivoted at A and opposed by the spring at B. A variation of this form places a hinge at P and the spring

FROGS, SWITCHES AND SWITCH-STANDS. 185

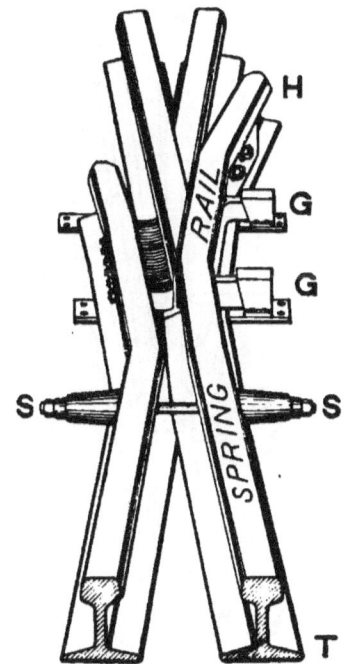

Fig. 202.—Typical Spring Rail Frog.

Spring-rail frogs.

nearer A, while still another kind following the same idea is exhibited in fig. 204. Within the last few years a small number of double spring-rail frogs have been built and these provide a continuous rail for both tracks. They are not much needed except in busy yards and at the entrances to terminal passenger stations.

In ordering frogs, the number (or angle) and the total length of the frog, the gage of the track, the section and drilling of the rail should be given, and if a spring-rail frog, whether a "right hand" or "left hand" is desired. This is determined by standing at the head of the switch and looking toward the frog. It will then be seen whether the frog is to go into the rail upon the right hand or into the rail upon the left hand. In short

Ordering frogs.

Spring-rail frogs.

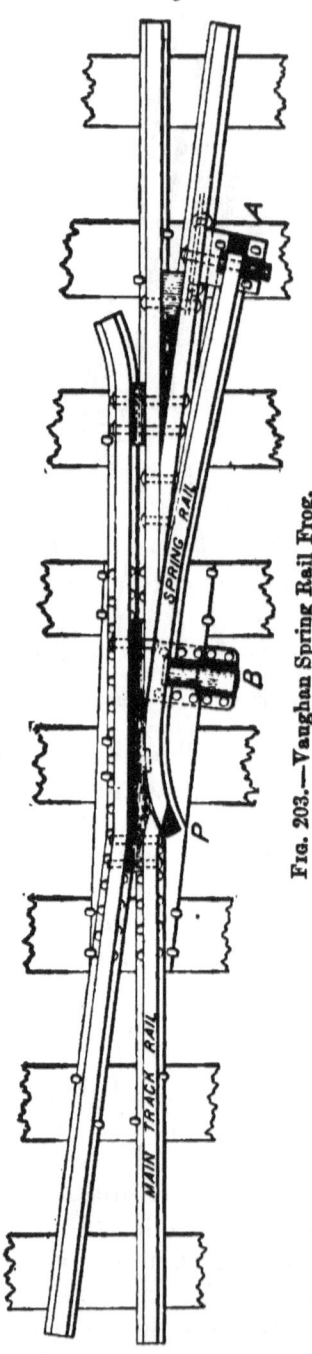

Fig. 203.—Vaughan Spring Rail Frog.

FROGS, SWITCHES AND SWITCH-STANDS. 187

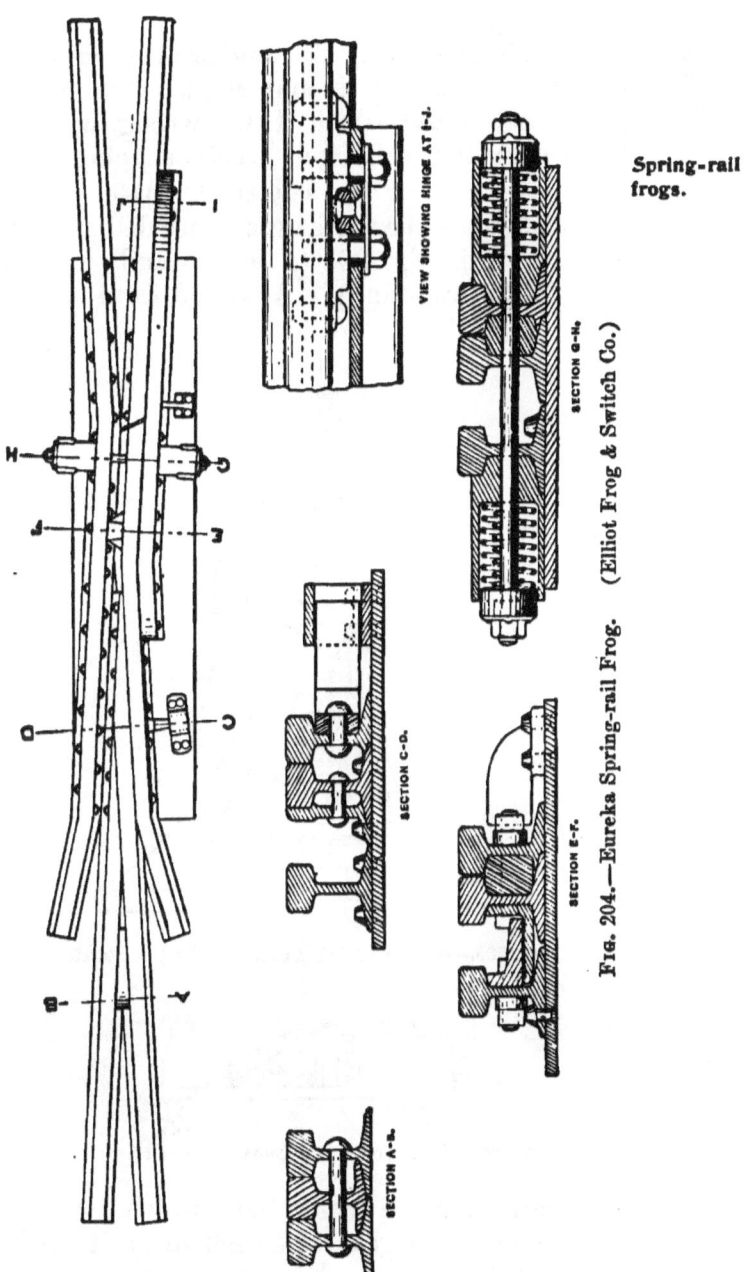

Spring-rail frogs.

FIG. 204.—Eureka Spring-rail Frog. (Elliot Frog & Switch Co.)

fig. 203, is a right hand and fig. 204, a left hand frog.

Automatic switch stands. Switch stands may be divided into two general classes, automatic and rigid. Automatic stands are those which if set wrong will be thrown by the train itself whether going over the main track or the side track route. Two of them, both operated by a concealed spring, are shown in fig. 205 (a high stand) and in fig. 206 (a low stand). A second form of low

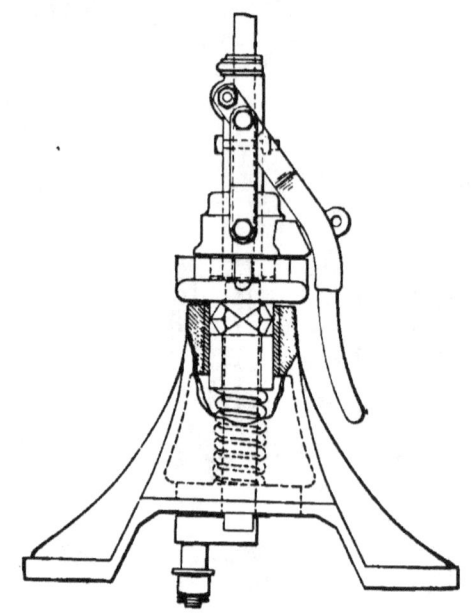

FIG. 205.—Ramapo High Automatic Switch Stand.

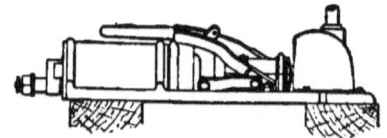

FIG. 206.—Ramapo Low Automatic Switch Stand.

automatic stand in which the throwing mechanism consists of gearing, is shown in fig. 207, while still another where the mechanism is also of gearing (somewhat differently arranged from that in fig. 207) is exhibited in fig. 208. Fig. 207, combined with a high target, is **Automatic switch stands.**

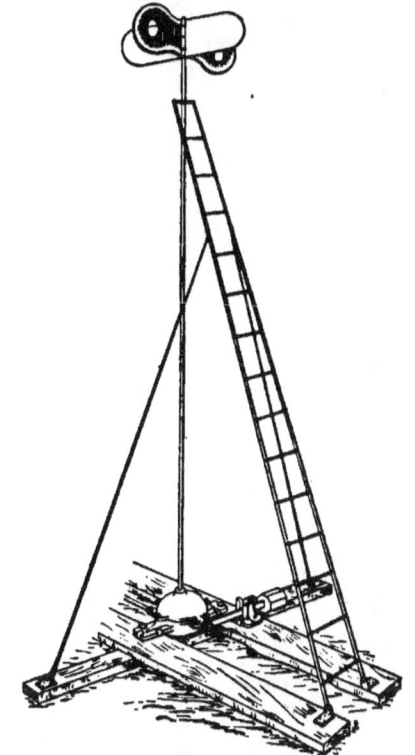

FIG. 207.—Pennsylvania Steel Co. Automatic Switch Stand and High Target.

FIG. 208.—Eclipse Automatic Switch Stand.

an extremely good arrangement, since it can be seen at a greater distance than low targets, and

Switch stands. will almost always distinguish main line from other switches. All of the previous stands are intended for single-throw split switches, but fig. 209 is to be used at three-throw split switches.

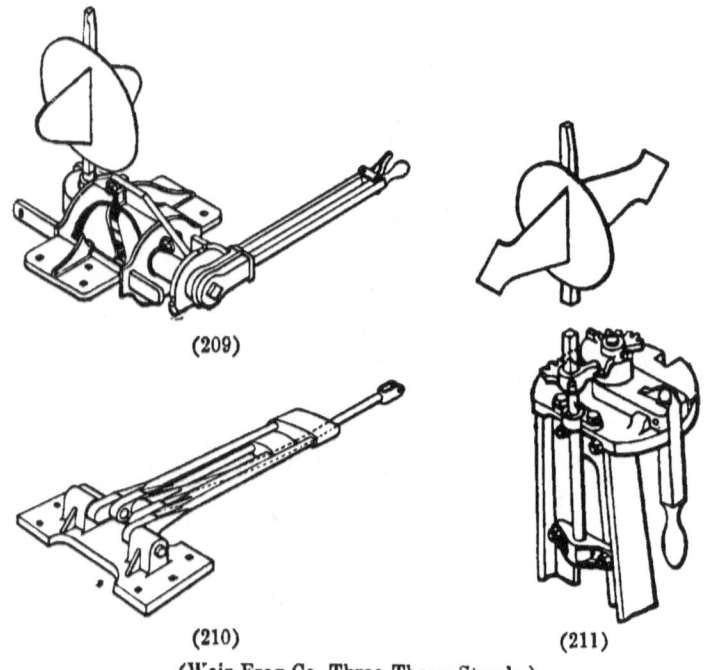

(209)

(210) (211)

(Weir Frog Co. Three-Throw Stands.)

FIG. 209.—Low, For Split Switches.
FIG. 210.—Low, For Stub Switches.
FIG. 211.—High, For Stub Switches, with Target-Throwing Attachment.

Rigid stands. Figs. 210 and 211 show ingenious stands especially intended for stub switches. Fig. 211 is to be particularly recommended since by means of the cog wheels on its top plate, the target is made to show red for both side tracks, no matter whether the main track is in the center or not. This is an impossibility with the ordinary three-throw switch stand. These target-moving cogs may also be applied to

fig. 209. The old reliable "jack knife" stand **Switch** which may be used anywhere and with any **stands.** target, single switch or movable frog, is illustrated in fig. 212. It is the best representative

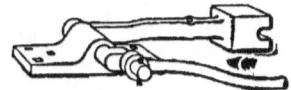

FIG. 212.—Jack-Knife Switch Stand.

of the rigid stand except at busy switches where deep snow is to be expected, at which points the harp stand, fig. 213, may be substituted.

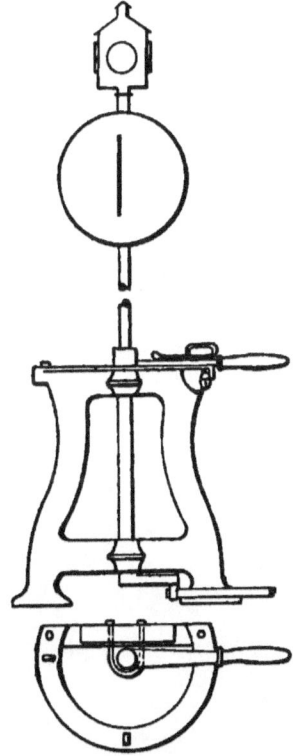

FIG. 213.—Harp Stand for Single or Three-Throw Switches.

The switch lamps and the signal lamps **Switch** should not be taken from the stands in the **lamps.**

Switch lamps. morning until the targets can be seen for a considerable distance. They should be cleaned and filled every day and the wick should be trimmed by rubbing, not by cutting it; finally the lamps should be lighted and allowed to stand for at least half an hour in the evening, before being taken out, so as to make it certain that they will not smoke.

On some roads it might appear to an observer that the switch lamps are put up more as a formality than for actual use. The writer believes that one of the most frequent causes for their going out is loose head blocks which are left untamped at the end under the switch stand and this is evidently an easy matter to correct. A lamp of bad construction or poor oil the trackman is of course not responsible for.

CHAPTER XIV.

EMERGENCIES AND TRAIN SIGNALS.

Since nearly everything which can happen to stop the traffic of a railroad will, in some particular, damage the permanent way, the presence of the trackmen will usually be required to make repairs. It follows, therefore, that they must be prepared at all times to turn out in full force with the tools and materials necessary for the work. The tool-house should be placed next to a track which is not used for standing cars. The tools of all kinds should be kept at hand and in good order while the men should live within easy call of the section-foreman.

Extra material. On each section, some rails, bolts, spikes, angle bars and cross ties should be stored in convenient and safe places to meet the sudden demand caused by train wrecks, floods, landslides, and so forth. At the subdivision headquarters there should be kept on hand some timbers of different lengths, preferably 12 in. square, since that is the most useful size.

Getting to the trouble. If the seat of trouble is not more than fifteen miles from the tool-house, the trackmen upon being notified should start at once on their hand-cars, rousing those gangs whose houses they pass and who are not located near a telegraph office. When a large force is required it will probably pay to start an engine and car to pick up the most distant gangs, unless the wrecking train must pass over that track.

Wrecking force.

Because most roads have a regularly organized wrecking crew made up from the car repairers or some other class of men who are familiar with the construction of rolling stock, it is not likely that the maintenance-of-way force will be called upon to do much independent work of this kind. Trackmasters and section-foremen nevertheless should familiarize themselves with the general features of clearing up wrecks of all kinds; how to move heavy weights, how to tie different kinds of knots, the best way of putting a derailed car on the track and the use of a block and tackle.

Knots.

Seven different kinds of knots are shown in fig. 214: A is a square knot which will not

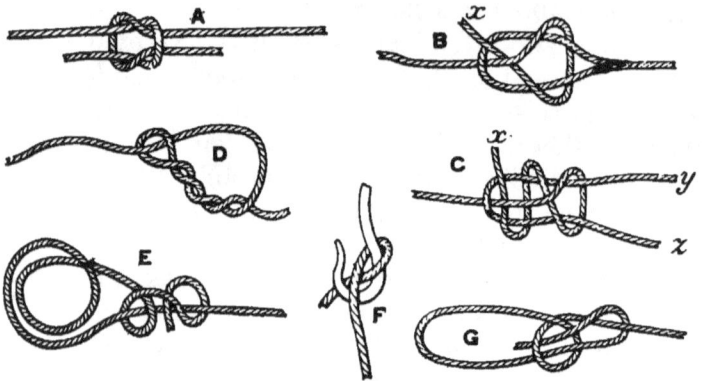

FIG. 214.—Rope Knots.

slip and is used in joining the ends of two ropes, but it is difficult to untie. Most persons attempting to tie a square knot fail, and make a "granny knot," which will slip. The failure is due to winding the short ends together in the reverse way. A few experiments with a string will teach one how not to do it. B and C are alike and for the same purpose, except that the rope x takes one more turn in C than it does in B. These knots may either of them be used with a loop, as in B, to

join the ends of two ropes or to fasten a rope **Knots.**
x to the middle of a rope y-z as in C. E and
G are two methods of fastening to a tree, post
or cross tie. E is a slip-noose, and G is a
bow-line loop, which does not slide. Both
will untie readily with a large rope. F is seen
to be for attaching the end of a rope to the
hook of a block. All of this will prove useful
knowledge, not only when trackmen are called
upon to do actual train-wrecking but in regular maintenance-of-way service as well.

Upon arriving at the place where traffic is **Duties at a wreck.**
stopped, the first thing to do is to make sure
that flagmen are so placed as to warn trains in
time to prevent any further trouble. At a
wreck, the trainmen themselves are expected
to perform this duty but that fact should not
prevent roadmasters and section-foremen from
seeing that the matter is receiving attention.
Safety is, in all questions connected with a railroad, the first consideration.

The procurance of material for repairs should **Use of material.**
receive early attention in order that the track
may be made ready for trains as fast as the
obstructions are removed. Material should
be used carefully at all times but in emergencies
it may be necessary to ignore many ordinary
ideas of cost for the sole object of putting the
track and road-bed into a condition for the
resumption of traffic. For instance, in the case
of a washout where it may be better to fill
the opening with crib work built of new ties
than to wait until dumping material can be
secured.

In the meantime those men who are not **Orderly behavior.**
doing flag duty, or absent after material,
should turn in and help in every possible way.
A ready obedience should be granted to the
person in authority at the time and no department jealousy should be allowed to interfere

Repairs to track.

with the work in hand. Shouting and swearing by the foremen, grumbling or shirking by the men should not be permitted. Everyone on the ground should work hard, cheerfully and, with the exception of those directing the work, in silence.

The track at a wreck should be roughly straightened as fast as the wreckage is cleared away but general repairs should not be attempted until the way is clear and no more car bodies or tenders are to be skidded on the rails or dragged along the ties.

Steel rails which have been distorted in any way by a sudden blow can never be safe for the main track unless cut and spliced at the bend, which is bad practice for it makes one more joint to keep up. They should therefore be taken out before fast traffic is resumed and replaced with sound metal.

The final repairs to a piece of damaged track should be made before withdrawing the men, even temporarily, if it is a possible thing. The track should be re-lined, re-surfaced, fully spiked and bolted, and if it is a train wreck which caused the damage, that part of the wreckage which is of no value, should be got out of the way immediately, in order to remove it from the public eye, which is quick to see and comment on such things.

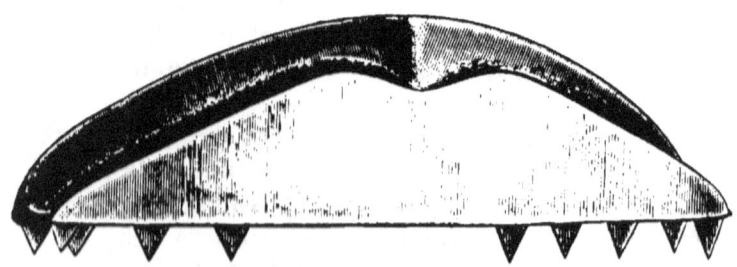

FIG. 215.—Alexander Car-Replacer.

On smaller railroads where there is no regular wrecking force, the trackmen will be forced to act in many cases which would ordinarily be out of their province. Under such conditions many tools will be required, not usually included in the maintenance-of-way list. Among the most important of these is a car-replacer, and one of many forms is shown in fig. 215.

Train Signals.

It is a necessary part of the equipment of a roadmaster, supervisor or section-foreman that he shall be acquainted with the rules governing the use of signals of all kinds and for this reason certain of the rules contained in the Standard Code are inserted here as follows:

All employees whose duties may require them to give signals must provide themselves with the proper appliances, and keep them in good order and always ready for immediate use. Flags of the proper color must be used by day, and lamps of the proper color by night or whenever from fog or other cause the day signals cannot be clearly seen. General Instructions.

Red signifies danger and is a signal to stop.

Green signifies caution and is a signal to go slowly.

White signifies safety and is a signal to go on.

Green and white is a signal to be used to stop trains at flag stations for passengers or freight.

Blue is a signal to be placed on a car or an engine to forbid its being moved.

A torpedo placed on the top of the rail, is a signal to be used in addition to the regular signals. Torpedoes and fusees.

The explosion of one torpedo is a signal to stop immediately; the explosion of two torpedoes not more than 200 feet apart is a signal to reduce speed immediately, and look out for a danger signal.

A fusee is a signal which may be used in addition to the torpedoes or other signals.

A flag or lamp swung across the track, a hat or any object waved violently by any person on the track, signifies danger and is a signal to stop.

Train flags and lamps. Each train, while running, must display two green flags by day and two green lights by night, one on each side of the rear of the train, as markers, to indicate the rear of the train. Yard engines will not display markers.

Each train running after sunset, or when obscured by fog or other cause, must display the head-light in front, and two or more red lights in the rear. Yard engines must display two green lights instead of red, except when provided with a head-light on both front and rear.

Two green flags by day and night and, in addition, two green lights by night, displayed in the places provided for that purpose on the front of an engine, denote that the train is followed by another train, running on the same schedule and entitled to the same time-table rights as the train carrying the signals.

Two white flags by day and night and, in addition, two white lights by night, displayed in the places provided for that purpose on the front of an engine, denote that the train is an extra. These signals must be displayed by all extra trains, but not by yard engines.

A blue flag by day and a blue light by night, placed on or at the end of a car, engine or train, denote that workmen are at work under or about the car, engine or train. The car, engine or train thus protected must not be coupled to or moved until the blue signal is removed by the person who placed it.

When a car, engine or train is protected by a blue signal, other cars must not be placed in front of it, so that the blue signal will be obscured, without first notifying the workman, that he may protect himself.

One *long* blast of the whistle is the signal for approaching stations, railroad crossings and junctions (thus, ——). **Whistle signals.**

One *short* blast of the whistle is the signal to apply the brakes—stop (thus, -).

Two *long* blasts of the whistle is the signal to throw off the brakes (thus, —— ——).

Two *short* blasts of the whistle is an answer to any signal, except "train parted" (thus, --).

Three *long* blasts of the whistle is a signal that the train has parted (thus, —— —— ——).

Three *short* blasts of the whistle, when the train is *standing*, is a signal that the train will back (thus, ---).

Four *long* blasts of the whistle (thus, —— —— —— ——) is the signal to call in a flagman from the west or south.

Four *long* followed by one *short* blast of the whistle (thus, —— —— —— —— -) is the signal to call in a flagman from the east or north.

Four *short* blasts of the whistle is the engineman's call for signals from switch-tenders, watchmen, trainmen and others (thus, ----).

Five *short* blasts of the whistle is a signal to the flagman to go back and protect the rear of the train (thus, -----).

One *long* followed by two *short* blasts of the whistle is a signal to be given by trains when displaying signals for a following train, to call the attention of trains to the signals displayed (thus, —— --).

Two *long* followed by two *short* blasts of the whistle is the signal for approaching road crossings at grade (thus, —— —— --).

A succession of *short* blasts of the whistle is an alarm for persons or cattle on the track, and calls the attention of trainmen to danger ahead.

<small>Hand and lamp signals.</small>

A lamp swung across the track is a signal to stop.

A lamp raised and lowered vertically is the signal to move ahead.

A lamp swung vertically in a circle across the track, when the train is *standing*, is the signal to move back.

A lamp swung vertically in a circle at arm's length across the track, when the train is *running*, is the signal that the train has parted.

A flag, or the hand, moved in any of the directions given above, will indicate the same signal as given by a lamp.

CHAPTER XV.

Fixed Signals.

The practice of placing fixed signals on the line of a railroad is becoming so general that trackmen should be acquainted with the significance and appearance of the most modern kinds. No attempt will be made in this chapter to explain many details of construction; for the maintenance of signal plants on most large railroads is now, and should remain, under a separate department.

All of the railroad signals with which we now deal are for the general purpose of maintaining a safe interval of space between moving trains, in order that they shall not collide, and this is effected in two ways. First, by interlocking signals, which relate solely to trains running upon separate but converging tracks. Second, by block signals (for description see page 224), which refer only to trains running upon the same track. The "signals" to be described, are devices located at fixed points, close to the line of a railroad, for telling the men in charge of a train whether or not the track they are upon is ready for their occupation beyond the point at which the signal is placed. These signals are said to "command," "govern" or "control" the movement of trains over the tracks to which they relate, and trainmen are said to be "governed" or "controlled" by the signals as they pass from one point to another on the tracks of a railroad.

Purpose of signals.

Interlocking. Interlocking signals are those which are made to work in connection with the shifting parts of a railroad track, such as movable frogs and switches. They are so arranged that, first, no train shall proceed until all of the tracks have been placed in their proper position; second, no train shall proceed until all other trains which might collide with it have been warned to stop; third, none of the shifting parts of a track can be moved so long as a signal gives the indication to proceed. The term "interlocking," therefore, refers to the relation which exists between the movable parts of a system of tracks and the signals which control the operation of trains through that system.

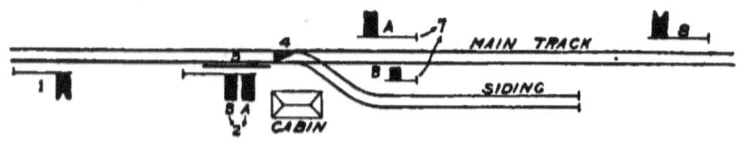

Fig. 216 —Single Track Joined by Side Track.

Names of signals. Fig. 216 is a conventional drawing of a single main track joined by a siding. It is sufficient now to give the names of the different parts, because their office and construction will be explained further on. In fig. 216 numbers 2A and B and 7A are *home signals;* 7B is a *dwarf signal;* 1 and 8 are *distant signals,* 4 is a *switch,* 5 is a *facing-point lock,* while numbers 3 and 6 are not used but are retained as *spare levers* in the machine, fig. 217.

Machine. All of these devices are operated by a collection of *levers* placed side by side in a common frame. This collection is called a *machine* and is located in a building conveniently situated (see fig. 216) known as a *cabin*. Forming a part of the machine are various pieces spoken of as "the interlocking" which, following the motion of the levers, interfere

Interlocking machine.

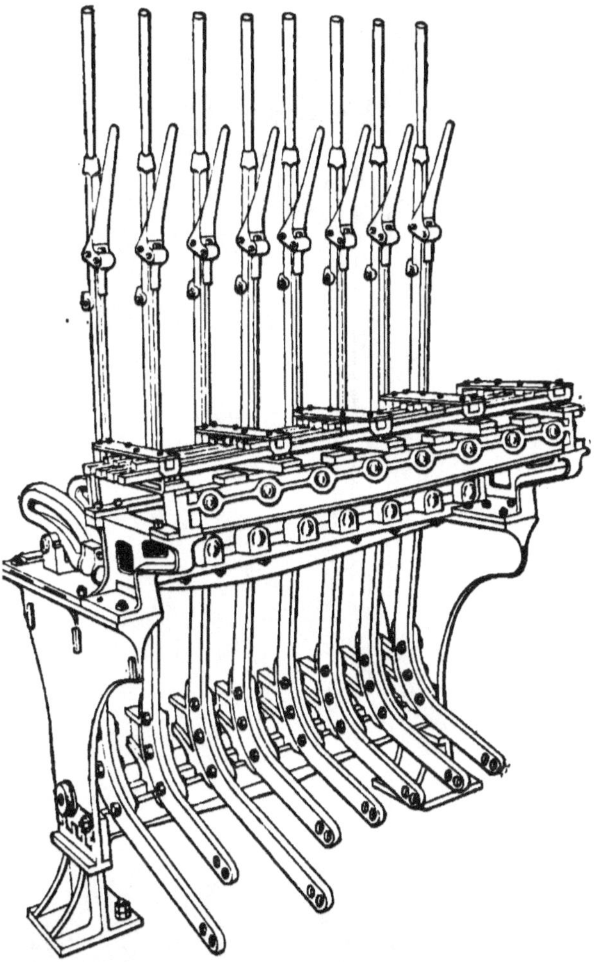

Fig. 217.—Interlocking Machine.

with each other after a certain manner and accomplish the purposes named in the foregoing paragraph entitled "Interlocking."

The signals used in interlocking are *semaphores*, that is *arms* (also called *blades*), projecting from a vertical post and so pivoted

Description of signals.

Home signal. to it as to be capable of swinging up and down. Semaphores are of three styles, *home*, *distant* and *dwarf*. The home signal, fig. 218,

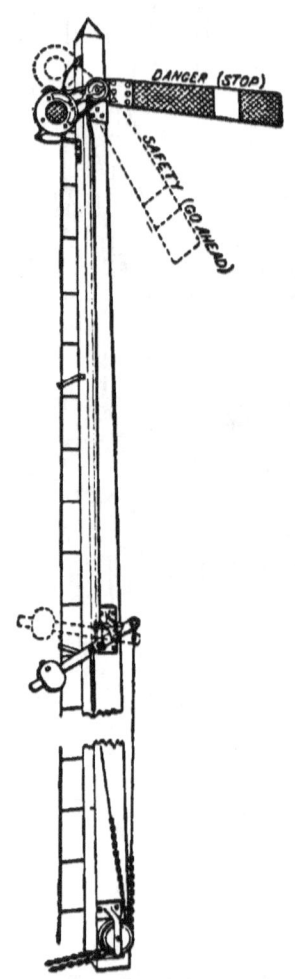

FIG. 218.—Home Signal.

is an arm about 5 ft. long by 8 in. wide with a square end, painted red (usually) on the face and white on the back; placed about 25 ft. above the rail and if possible to the right of the track which it controls. When more than one arm is placed on a post, fig. 216, No. 2, the upper arm is for the most important track (sometimes called "route" in this connection) and the lower arm is for all other routes which connect with that track and at the same time come under the control of that signal. The home signal is used only for movements on main track. On double track, fig. 219, the home signal is used in only one direction for each track. On single track, the home signal

Home signal.

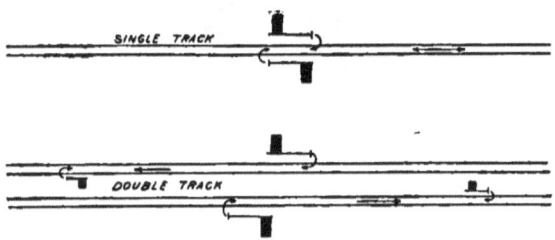

Fig. 219.—Home Signals on Double and Single Track.

is used for both directions, since trains are run in both directions. To indicate danger, see fig. 218, the arm stands horizontally and shows a red light (usually) at night; to indicate safety the arm is inclined about 65 deg. from the horizontal and shows a white light (usually) at night.

The distant signal, fig. 220, is an arm about 5 ft. long by 8 in. wide with a notched end, painted green (usually) on the face and white on the back; placed at the same height as the home signal, about 1,500 feet away from it, and on that side of the home signal first reached by the trains which it governs. It is used only in conjunction with some particular arm of a home signal, never alone, and merely for the purpose of warning enginemen as to the probable position of that home signal. It indicates either caution (go slowly) or safety by the positions shown in fig. 220 in the day time. At night, caution is indicated by a green

Distant signal.

Distant signal. light (usually) and safety by a white light (usually).

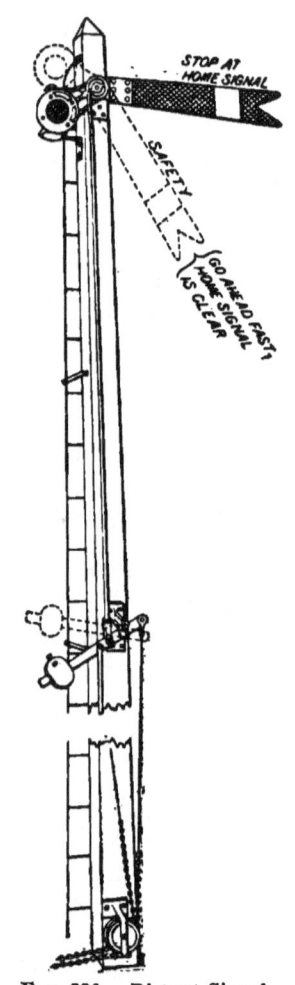

Fig. 220.—Distant Signal.

Dwarf signal. The dwarf signal, fig. 221, is a blade about 1 ft. long, with a square end, painted red (usually) on the face and white on the back. It is placed about 3 ft. above the rail and usually to the right of the track that it governs. It is used only for train movements against the usual direction of the traffic on double main track, but never on single main track, see **Dwarf signal.** fig. 219, and it is used to control all move-

FIG. 221.—Dwarf Signal.

ments in any direction on "side" tracks. The dwarf signal gives its indications by the same relative positions and lights that are used by the home signal. It seldom carries more than one arm and this one governs all routes over which it has control.

On almost all railroads in this country, sem- **Pointing of arms.** aphore arms point to the right when viewed from approaching trains which they govern, and, although a semaphore post may carry arms which govern trains moving in opposite directions, as in fig. 222, a certain arm never governs trains moving in opposite directions.

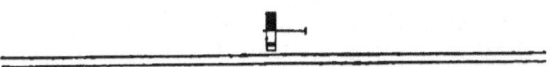

FIG. 222.—Semaphore controlling trains from both directions.

No two home signals which are located on the same post, as in fig. 216, No. 2, can be lowered at the same time.

No ordinary semaphore posts, such as are **Bracket posts.** shown in fig. 216, Nos. 1, 2A and B, 7A and

Bracket posts.

B, ever control trains moving on more than one track. When it is necessary to do so a special post called a "bracket-post" is provided, see fig. 223, where a siding stands so

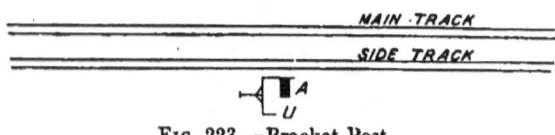

Fig. 223.—Bracket Post.

close to a main track that it is impossible to place a post between them. The arm A therefore controls trains moving to the right on the main track and the "dummy" upright U, which carries a blue light at night, indicates that the side track is not signalled.

Movement of trains.

In fig. 216 all of the interlocked parts are shown in what is called their "normal" position. This corresponds to the forward position of the levers in the machine. It will be noted that the signals are all at danger and the switch is set for the main track as indicated by the flare, thus:

If the switch were set normally for the side track, a thing often done, it would be indicated by a flare thus:

In the case of the facing-point lock, No. 5, fig. 224, since the switch is always unlocked

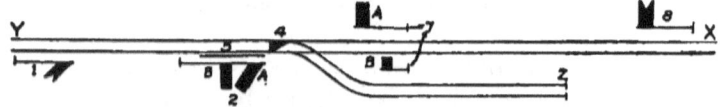

Fig. 224.—Tracks and Signals.

when the facing-point lock lever is "normal" (that is "forward") and unlocked when the facing-point lock lever is "reversed," or in other words "pulled back," the position of the facing-point lock needs no other identification on the drawing than to merely show its presence.

Movement of trains.

If now a train bound for X were to approach from Y, it should find the signals set as in fig. 224, which, because the distant signal 1 and the home signal 2A are "inclined" (also expressed as "cleared," "dropped," "lowered") would indicate that the main track route had been cleared throughout the system. A train bound from Y to Z would find the signals and switch as shown in fig. 225. Here the bottom arm B

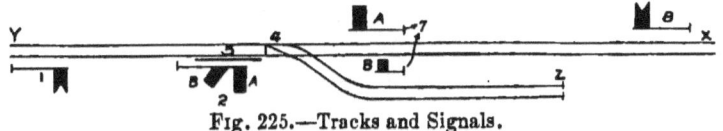

Fig. 225.—Tracks and Signals.

on home signal post, 2, must be lowered because the switch has been prepared for a "diverging route," that is a route which would carry trains away from the most important track. The distant signal 1, must remain at caution for it cannot be cleared until the home signal, 2A, has been previously cleared. In both of the cases illustrated in figs. 224 and 225 the facing-point lock, 5, must have been reversed before any of the signals could have been cleared, and this together with the relations between home signal, 2A, and distant signal, 1, are made unavoidable by means of the "interlocking" feature of the machine mentioned in the beginning of this chapter. A train coming from X, in fig. 226, has but

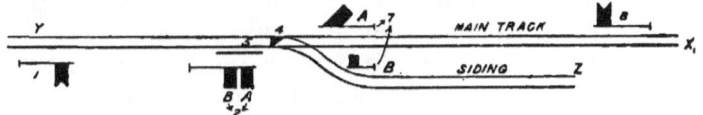

Fig. 226.—Tracks and Signals.

Movement of trains. one route possible and this is indicated as clear when the train reaches home signal 7A. It must have approached at a slow speed however for distant signal, 8, was found at caution. A distant signal cannot be cleared until after the home signal with which it works has been cleared, but there is nothing to force the clearing of the distant signal at all unless the signal-man wishes it so. The last combination possible with the tracks shown in figs. 224, 225, 226 and 227, is exhibited in

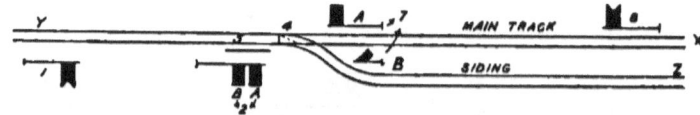

Fig. 227.—Tracks and Signals.

fig. 227, and contemplates the movement of a train from Z to Y. In this case switch, 4, has been previously reversed and the dwarf signal, 7B, has alone been cleared. In each case the clearing of any home signal has locked fast all of the other home and dwarf signal levers in the cabin, for it is evident that if 7A and 7B were cleared at the same time a collision might result. The same is equally true of the relations between 2A and B and 7A and B.

Placing of signal posts. A particular meaning is attached to the way in which signals are shown on a plan, and this is further explained in fig. 228. Ordinarily signals B 1 – 2 would stand at A in the form of a straight two-arm post or at H on a bracket-post shaped like D, but the first is impossible because the space between the tracks is assumed to be insufficient, and the second because of the freight-house which stands in the way. The arms are therefore placed on a bracket-post, pointing to the left, and are shown as white with black bands, which is the appearance they present when seen from an approaching train **Placing** which they do not govern, as would be the **of signal** case if a train went towards them from 5. **posts.** The function of signals C, D, E and K is

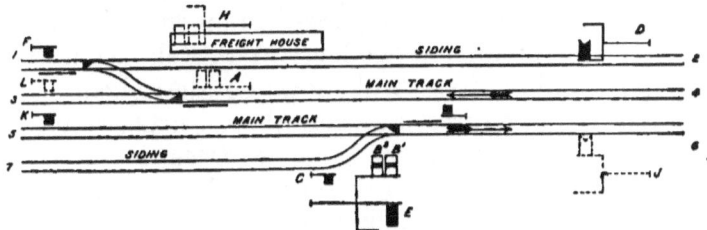

FIG. 228.—Placing of Signal Posts.

plain from reasons before stated. D is evidently the distant signal for B', and it is placed at D rather than at J, because if possible it is preferred to have all signals on the right of the tracks which they govern when viewed from an approaching train. This is not possible (we assume) at L, so the dwarf signal for the "crossover" is placed at F, that is on the left of the track, 1 – 2, but with its blade pointing to the right and marked in black, showing that it controls trains moving from 1 to 2 or 1 to 4.

The simple "split switch," the "derail," the **Combina-** "double-slip" with or without the "movable **tions of** frog," and the "single-slip" with or without the **switches.** "movable frog," are all used in connection with interlocking machinery, and the way in which they are indicated on a plan is shown in fig. 229. Here $1^1 - 4^2 - 6^1$ are simple "split

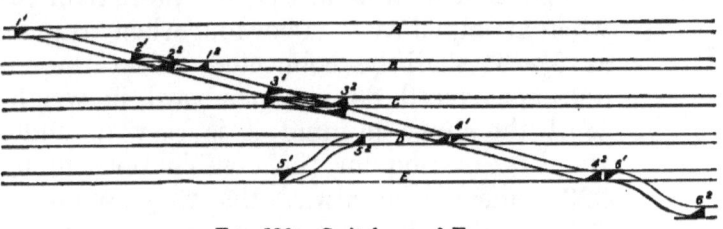

FIG. 229.—Switches and Frogs.

Combinations of switches.

switches," $2^1 - 1^2 - 2^2$ is a "single-slip with movable frog," $3^1 - 3^2$ is a "double-slip with rigid frog," 4^1 is a "movable frog," 6^2 is a "derail," and $5^1 - 5^2$ an ordinary crossover formed of two simple split switches. Certain combinations of these arrangements may be operated from one lever in a machine and those combinations will now be described. The large figures indicate those parts which are worked from the same lever as $1^1 - 1^2$, while the small figures serve only to distinguish them from each other. All of the derails, frogs and switches are shown in their normal positions and are marked with the numbers of the levers which operate them. The small figures at the top of each number are for reference here and are not used in practice. The "crossover" $1^1 - 1^2$ is now set so that a train on tracks A or B would follow the straight route, but if set like $5^1 - 5^2$, a train would be forced to go from one track to the other; therefore both switches may be worked from the same lever, 1. This is also the case with 1^1 and 4^2, which might properly be worked together except for the fact that they would cut off all traffic on tracks B, C, D and E during the time they might be reversed. There is of course an inevitable limit to the number of switches that may be operated from one lever, which is determined by the amount of power which may be applied, and experience dictates, when switches are to be operated by a man, that not more than two when arranged as $1^1 - 1^2$ or four when arranged as $3^1 - 3^2$ shall be connected, when the rail does not exceed 80 lbs. per yard in weight. With the above in mind it is easy to understand the reason for the combinations in fig. 229, remembering always that they are not the only ones possible but are given because they are proper and serve the purpose of explanation.

Facing-point locks.

The facing-point lock, fig. 230 (indicated as No. 5 in fig. 216), consists of a casting bolted

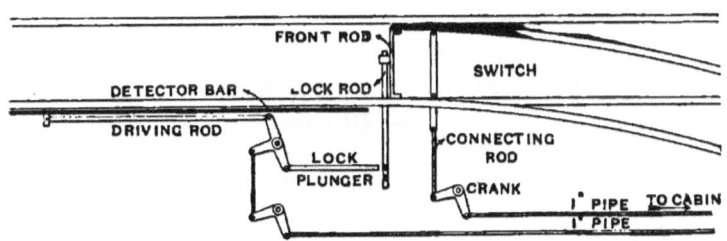

FIG. 230.—Facing-Point Lock.

to the ties in front of the point of a switch, through which slide the lock-plunger and lock-rod at right angles to each other. The lock-plunger is connected directly with a lever of the machine in the cabin by means of cranks and 1-in. pipe, while the lock-rod is also connected with the machine but receives its motion through the switch, to the front rod of which it is fastened. A complete movement of the switch brings one or other of the openings in the lock-rod into place in front of the lock-plunger, which, when it is reversed by its lever in the machine, locks the switch fast. But if the switch be not given its full travel, the lock-plunger will impinge against the solid metal of the lock-rod. The lock-plunger lever will thus be prevented from completing its journey and, the interlocking parts of the machine being in consequence out of place, it will become impossible to clear the home signal. The facing-point lock therefore has two duties: first, to lock the switch if it is in the proper position and, second, to prevent the clearing of the home signal if the switch happens to be wrong.

Detector bar.

A detector-bar is almost invariably used with a facing-point lock and is assumed to be present unless it is specifically stated to the contrary. This device is illustrated in fig. 231

 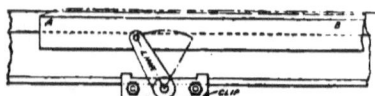

Fig. 231.—Detector Bar.

and is seen (fig. 230) to be connected with the same lever in the cabin which operates the facing-point lock. It is to prevent the unlocking of a switch while a car is standing over (straddling) a switch. This is necessary since any movement of a switch at such a time might result in a derailment. The bar is a piece of iron or steel usually about 45 ft. long, (which distance is expected to be greater than the greatest distance between any two wheels of a train), extending from the point of the switch as shown in fig. 230. In this arrangement only one bar is necessary, but in fig. 232, where it is back of the head-block of the

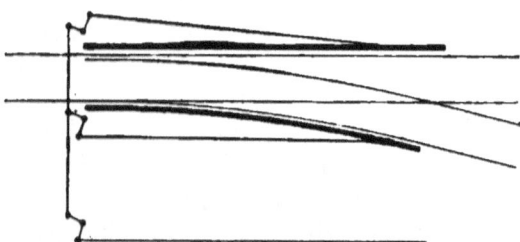

Fig. 232.—Special Arrangement of Detector Bars.

switch two bars must be used, since the train may be standing on either of the two tracks. The detector-bar, fig. 231, is placed close against the rail with its top A – B, fig. 231, normally about $\frac{1}{2}$ in. below the top and us-

ually on the outside of the rail. It is sup- **Detector bar.**
ported on links pivoted at their bases in such
manner as to force the bar to rise about an
inch as indicated by the dotted line above the
rail, when it is moved back and forth by the
detector-bar driving-rod. By so rising above
the rail, the bar strikes any wheel which might
be standing there and because it is unable to
follow its full course, prevents the facing-point
lock from being withdrawn.

On side-track switches and on trailing-point **Switch and lock movement.**
main-track switches, a device, fig. 233, called

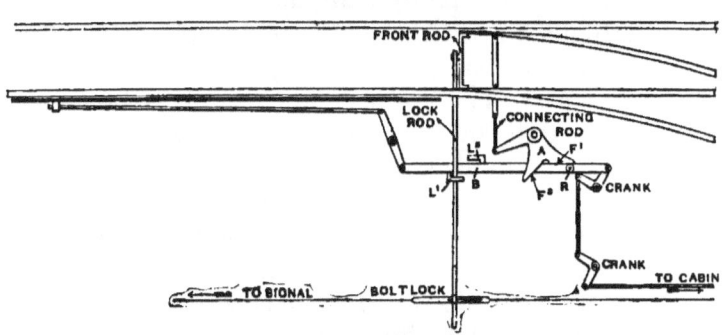

Fig. 233.—Switch and Lock Movement.

a "switch-and-lock-movement" is often used
for combining the operation of a switch, a
lock and a detector-bar from one lever, instead
of dividing it between two levers as in fig. 230.
It is an inferior method but is cheaper and is
only proper where the speed of trains is uniformly slow. The alligator-crank, A, and the
slide-bar, B, are mounted upon the same base.
To the arm of A is fastened the connecting
rod of the switch, while to the slide-bar, B,
are fastened the lock pins, L, the roller, R,
the driving-rod of the detector-bar and the
pipe connection to the cabin. As fig. 233 is
shown, the lever in the cabin is in one of its
extreme positions, the switch is set for the main
track and is held there by the lock pin, L^1, which

Switch and lock movement

is seen projecting through the lock rod. Upon moving the lever in the cabin, B is pushed to the left, which immediately operates the detector-bar and L^1 is withdrawn. In the meantime R is sliding along the face, F^1 of the alligator-crank, but no movement in the switch takes place until R reaches the face F^2. By this time the lock, L^1, is entirely clear and the detector-bar has reached its highest position above the rail. Then R forces A around until F^2 is parallel with B. This corresponds with the new position of the switch and takes place just before L^2 enters the opening in the lock rod, which together with the complete lowering of the detector-bar is the last operation of the movement. The same sequence of action takes place in the contrary course of B.

Bolt lock.

A "bolt-lock," fig. 233, is sometimes used as a check on the action of a switch-and-lock-movement (occasionally elsewhere) and consists of a rod (the extension of the lock-rod) and a bolt-lock. Its purpose is to prevent the clearing of a signal should the switch not be exactly right. To each end of the bolt-lock is connected the wire which joins the signal to its lever in the cabin. In fig. 233, the signal has been cleared because the bolt-lock is seen to have entered the opening in the lock-rod, but if the opening had not stood directly opposite the bolt-lock, that piece would have impinged against the solid metal of the rod and its further movement have been stopped, while as a result the signal would have remained at danger.

Selector.

It remains to mention but one more of the devices used in interlocking, before proceeding to a description of the machine, and that one is the "selector." Its object is to reduce the number of levers in a machine by enabling two or more signals to be operated from the

FIXED SIGNALS. 217

same lever and its essential parts are illustrated in fig. 234. Theoretically, any number

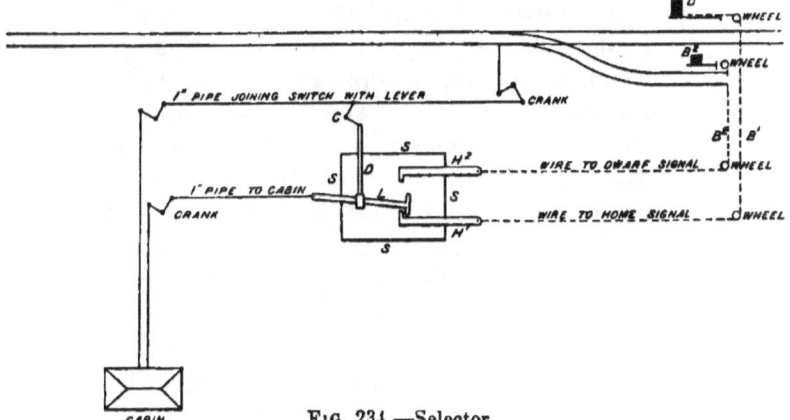

FIG. 234.—Selector.

of signals which govern trains moving in the same direction may be operated from a certain lever of a machine, if but one of those signals can be properly cleared at the same time. Practically not more than seven or eight signals are ever operated from the same lever. A selector is always described according to the number of switches in connection with which it works and not with regard to the number of signals, for in fig. 235, although switches 1

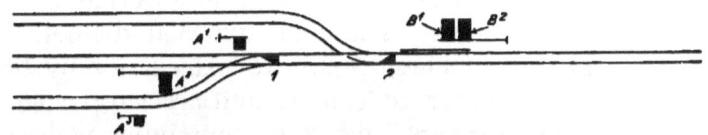

FIG. 235.—The "Selection" of Signals.

and 2 regulate the selectors for 3 signals (A) in one direction and for two signals (B) in the other direction, yet a "2-way selector" is used in each case. This is so because although there are three routes, signal B^2 is cleared whichever of the switches 1 or 2 is set for the divergent route. In consequence of this rule the selector of fig. 234, is a "1-way." A box,

Selector. S, fig. 234, contains two hooks, H, (three in a two-way, four in a three-way, etc.) which form the connection with the signal wires, a lug, L, (only one lug is used for all "ways") and a driving rod, D, (a separate driving rod is used for each switch, and the number in a selector therefore corresponds exactly with the size, 1-way, 2-way, etc., of the selector). The crank, C, connects D with the line of pipe which joins the switch with its lever in the machine, and D, consequently acts in accordance with the motion of the switch lever. Through a hole in D, L is loosely passed so that although it is moved by D laterally, nothing prevents the longitudinal motion of L. As the drawing is made, the switch stands for the main track and L is in connection with H^1 (the main-track signal hook). If the switch should be reversed by moving its pipe to the left, D would force L into connection with H^2 and L then being moved to the left would result in lowering B^2.

Pipe connections. Switches, facing-point-locks, detector-bars and switch-and-lock-movements together with a few other special devices, should be operated always, where man-power is used, by iron or steel, seamless pipe, having an internal diameter of one inch and an external diameter of $1\frac{1}{4}$ inch. These pipes are placed side by side, $2\frac{3}{4}$ in. center to center, and are supported in "pipe-carriers," fig. 236, containing rollers at

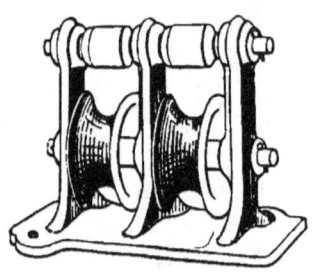

the top and bottom which confine the pipe and reduce the resistance to its motion. Changes in the direction of a pipe line are made (usually) by "bell-cranks," fig. 237, which rest

Bell crank.

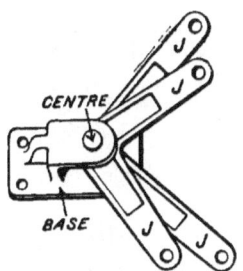

Fig. 237.—Bell Crank.

upon the base and revolve about the center. At the ends of the arms, J, the pipe is connected by means of "jaws," fig. 238, which

Jaws.

Fig. 238.—Jaw.

constitute the common method of attaching a pipe to some other article. The pipe is passed over the "tang" and is screwed into the sleeve, S.

Signals only, should be operated by wire and this should be of No. 9 galvanized steel, supported in "wire-carriers," fig. 239, which

Wire connections.

Fig. 239.—Wire Carrier.

are provided with rollers for the wire to rest upon. The changes in the direction of a line

Wire connections. of wire are made by inserting into the line a piece of ¼-in. chain with short links, and passing the chain around a wheel, fig. 240, which has a fixed center.

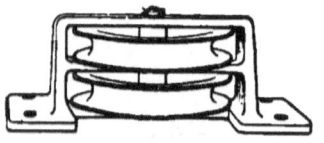

Fig. 240.—Chain Wheel.

Adjustment. Both pipe and wire will vary in length as the temperature changes and through other causes; they should therefore be provided with turnbuckles, fig. 241, to provide for small

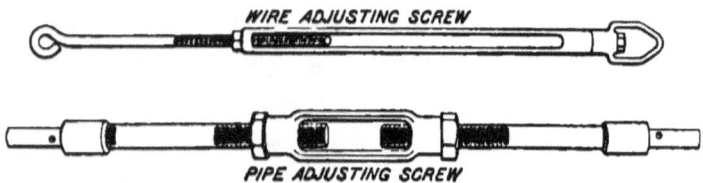

Fig. 241.—Pipe and Wire Adjusting Screws.

Compensators. adjustments. It is desirable also that they shall be automatically compensated but up to the present time, no perfectly satisfactory automatic wire-compensator has been devised, although there seems to be every reason to hope for one. The ordinary pipe-compensator, the "lazy-jack," fig. 242, is eminently successfull.

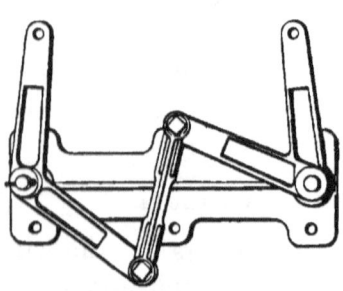

Fig. 242.—Pipe-Compensator ("Lazy-Jack").

FIXED SIGNALS. 221

The machine, figs. 217 and 243, consists of a frame to support the other parts, a series of levers for operating the different switches, signals, etc., and the interlocking mechanism which permits the movement of a lever at

Interlocking machine.

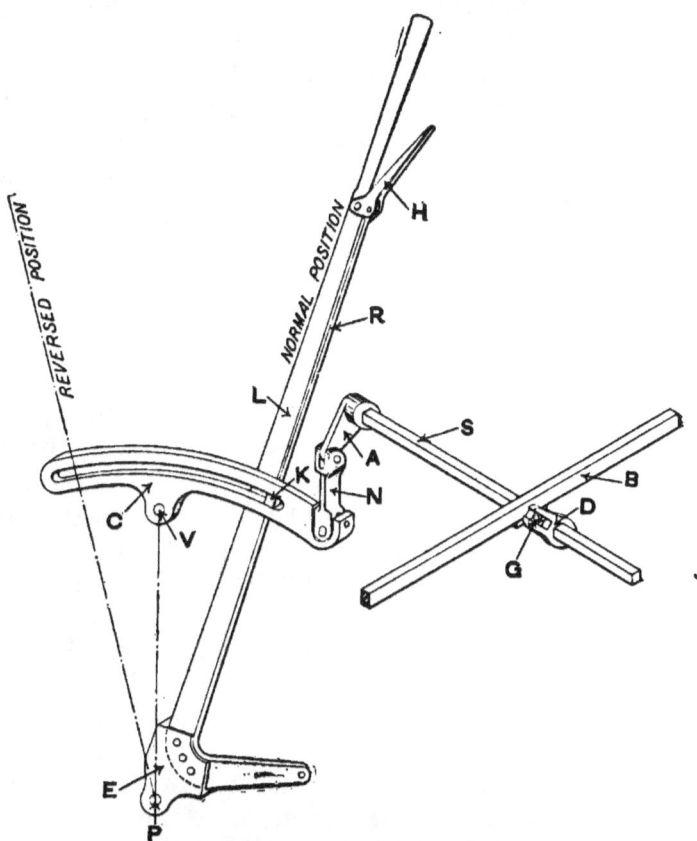

FIG. 243.—Interlocking Lever and Mechanism.

certain times. In fig. 243, L is one of a series of levers, E its lever-shoe, H a latch-handle, R a latch-rod, K a latch-block, C a rocker, N a rocker-link, S a locking-shaft and B a locking-bar, all belonging to that lever. In any machine a series of levers are placed 5 in. center to center, side by side in a frame. Each L is fastened to an E which is supported

Interlocking machine.

in the frame by a horizontal pin at P. L is shown in the drawing in the normal position; when reversed it coincides with the dotted line. The signals, switches, etc., are joined to the short arm of L at its lower end. H is pivoted at its lower end to L and when grasped by the hand, raises R, which in turn raises C by means of the block K. C is pivoted to the frame at V so that, when raised at K by R, it assumes a position circumferential to P; since it is formed on a curve whose radius is equal to K–P, it permits K to move through it freely as L is thrown forward and back. When L has been reversed, H is released, R is lowered and C assumes a third position. When L is normal, the right-hand end of C is depressed. During the movement of L, the two ends of C are at the same height, while when L is reversed and H released, the left-hand end of C is depressed. As L passes backward and forward between its two positions, K passes over a stop which prevents it from being lowered during any movement of L.

The vertical action of C causes N to be raised or lowered (depending upon whether L is to be moved from the normal or reversed position) which since N is connected with the crank-arm, A of the horizontal locking-shaft S, transmits the rise and fall of N to S, in the form of a rotary motion. Mounted upon B is a filling block, G, fitting into the driver, D, which is fastened rigidly to S. The rotary motion of S is thus transferred to a horizontal movement in B, while the final relation between the latch H and the locking-bar B, is completed as well as the way in which the three positions of C are communicated to B. From the preceding it follows that when L is changed from the "normal," B first moves to the left, then stops and finally completes its movement to the left. The contrary takes place when L is changed from the "reverse." The interlocking parts are mounted upon and are operated directly by the locking bars B, which are arranged in such a way that by the movement of any lever, all other levers are locked fast which if moved might in any way interfere with the train which it is intended to signal. *[Interlocking machine.]*

The details of the "interlocking" proper are however, too complicated to permit of explanation in a book not solely devoted to that subject and it must be taken for granted that the objects which are sought are successfully accomplished.

The principal devices used in manual interlocking have now been described and since that is the sole intention, a mention only will be made of other methods of operation. These are few. Man power it is believed must always remain the usual method of accomplishing the combined action of switches and signals, since it is the simplest and most easily controlled, so far, of all the forces at our disposal and must in any event be the means of intelligently controlling those forces, since it is self-evident that "interlocking" cannot become automatic. Hydraulic pressure for moving switches and signals has been tried and abandoned for many reasons. At the present time the field of power-interlocking is monopolized by what is known as the "Westinghouse Electro-Pneumatic Interlocking System" which indicates its character. All switches and signals are moved by compressed air conveyed to them by pipes from an air compressor. The valves which direct the action of the air are controlled by electromagnets. The machine is most ingenious, combining in itself all of the usual interlock- *[Power Interlocking.]*

ing features, together with certain electrical checks on the mutual operation of the switches, signals and machine levers. The system finds its best field at large installations where it is singularly successful through the great rapidity with which the changes in combinations may be made and because of the few leover-men who are required.

BLOCK SIGNALING.

Block-signaling. It will be remembered that in the definition of "interlocking" it was stated that the object of that branch of signaling is to preserve from collision, trains which are running upon separate but converging tracks; that is, tracks which either cross each other or join each other through the medium of a switch. In Block-Signaling the problem is quite different since its object concerns only those trains which are moving upon the same track and this includes both single track, where trains may be either approaching or following each other, and double track where trains only follow each other. To accomplish the separation of trains, a railroad is divided into sections of approximately equal length, called "blocks," with a signal placed at the beginning of each block. When a block is occupied, its signal should be in the danger position and when a block is empty, its signal may be in the clear position and a train may enter. This is block-signaling pure and simple.

Block signals. The signals used in block-signaling are preferably of the same appearance and meaning as those which have been described and are illustrated by figs. 218 and 220. There is consequently no reason for describing them a second time since every statement made concerning their functions in interlocking applies

equally to block-signaling. Until quite re- **Block**
cently "semaphore" signals, except in rare **signals.**
and unimportant instances, (the term used to
describe signals having an "arm") have only
been used where compressed air could be
applied to operate them, as in the Westinghouse automatic system, or where they could
be directly moved by the signal-man in a
cabin. But a few months before the issue of
this book, purely automatic semaphores have
been tried which derived their motion from an
electric current. These have worked with
encouraging success and if they finally inspire
confidence they will probably become the prevailing form of automatic signal.

FIG. 244.—The Banjo Signal.
(The Hall Signal Co.)

The "banjo" signal, fig. 244, is operated **Banjo**
wholly by electricity and consists of a trans- **signal.**
parent colored screen, enclosed in a case,
which shows through the opening of the case
when the signal is at danger (or caution for a

Banjo signal. distant signal) and is withdrawn from sight when the signal indicates safety. A lamp which shines through the small, upper opening in the case illuminates the signal at night.

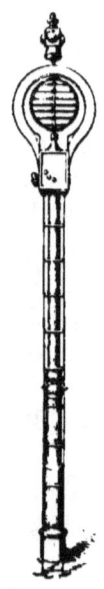

FIG. 245.—The Banner Signal.
(The Union Switch & Signal Co.)

Banner signal. The "banner" signal, fig. 245, is operated by a weight which is wound up periodically and is controlled by clock-work. This is alternately held and released by an electric current which operates a governing device, contained in the box upon which the signal rests. The signal revolves on a vertical shaft and exposes to the view of an approaching train, a banner of one form for danger and another form for safety. At night the signal indications are given by a revolving lamp, having lenses of different colors; this is mounted on the banner-shaft and therefore follows exactly the movements of the banners.

Track circuit. Nearly all purely automatic block-signals are controlled by what is known as the "track curcuit." This consists in having the rails in each "block" on the same side of the track, **Track circuit.** connected with each other electrically by short pieces of wire. The blocks are electrically separated by placing insulating material between the end rails and angle-bars of adjacent blocks. At one end of each block the last rails on the two sides of a track are joined by a wire in which an electric battery is placed, and at the other end of each block is placed a wire containing a track relay which controls the signal governing that block. This establishes an electric circuit which operates in such a way that, when a train occupies a certain block, the signal governing that block is cut out and forced to exhibit danger because the electric current is cut out from it by the presence of the train on the track. A clear (empty) block consequently results in a clear signal. Switches are also included in the track circuit so as to cause the signal to show danger in case the switch is not set and locked for the main track.

In all work about a railroad, trackmen must **Care of** be especially chary of changing in any way **appliances.** the operations or material of an interlocking or block signal plant. A mistake in this matter, such as the breaking of a piece of wire, may result not only in delays and inconvenience, which will be troublesome, but even in the loss of life. Repairs which require the movement or temporary abandonment of any signaling material should if possible be made under the direction of a man connected with the signal force and a sufficient notice should therefore be given whenever any such work is contemplated. Ignorant interference in signaling matters is more apt to result disastrously than in most other branches of railroad affairs and it should for that reason be more sedulously avoided.

CHAPTER XVI.

Rules and Tables.

Spikes. Railroad spikes are usually packed in kegs weighing 150 lbs. or 200 lbs. each. The spikes of common size, $5\frac{1}{2}$ in. long by $\frac{9}{16}$ in. square, run 280 to the 150 lb. keg. In other words, each spike weighs a little more than $\frac{1}{2}$ lb.

Bolts. Track bolts are more frequently packed in kegs containing a certain number of bolts rather than a certain number of pounds. The ordinary bolt with its nut weighs something less than 1 lb.

Rails. Rails are always sold by the "gross ton" which weighs 2,240 lbs., as distinguished from the "net ton" which weighs 2,000 lbs.

Cross ties. 7 in. by 9 in. by $8\frac{1}{2}$ ft. sawed, white oak, cross ties weigh about 195 lbs. each; 6 in. by 8 in. by $8\frac{1}{2}$ ft. sawed, white oak, cross ties weigh about 150 lbs. each. Hewed ties of the same classes weigh considerably more than the amounts given. Ties when purchased at a distance are received loaded on cars, which may be expected to contain from 150 to 250 ties each.

Capacity of cars. An ordinary flat car 33 ft. long by 8 ft. wide, with temporary sides 1 ft. high will carry about 18 cubic yards of loose material without spilling; without sides about 8 cubic yards. A gondola 33 ft. long by 8 ft. wide by 3 ft. 4 in. high, when loaded full but not heaped up, will carry about 32 cubic yards.

TABLE I.—Amount of Material Required for Different Lengths of Single Track.

Miles of Single Track			¼	½	1	2	3	4	5
Number of 30-ft. Rails			88	176	352	704	1,056	1,408	1,760
Gross Tons of Rails, Pounds per Yard.		65	25.5	51.1	102.1	204.3	306.4	408.6	510.7
		70	27.5	55.0	110.0	220.0	330.0	440.0	550.0
		75	29.5	58.9	117.9	235.7	353.6	471.4	589.3
		80	31.4	62.9	125.7	251.4	377.1	502.9	628.6
		85	33.4	66.8	133.6	267.2	400.8	534.4	668.0
		90	35.4	70.8	141.5	283.0	424.5	566.0	707.5
		95	37.3	74.7	149.3	298.6	447.9	597.2	746.5
		100	39.3	78.6	157.2	314.4	471.6	628.8	786.0
Pairs of Angle-bars			88	176	352	704	1,056	1,408	1,760
Number of Bolts, Nuts and Washers:	4-Hole Angle-Bars		352	704	1,408	2,816	4,224	5,632	7,040
	6-Hole Angle-Bars		528	1,056	2,112	4,224	6,336	8,448	10,560
Spikes and Cross-Ties (14 Cross-Ties to the 30-ft. Rail):	Cross-Ties		616	1,232	2,464	4,928	7,392	9,856	12,320
	Spikes		2,464	4,928	9,856	19,712	29,568	39,424	49,280
Cubic Yards of Ballast according to Fig. 66, Chapter VIII. Sawed Cross-Ties 7″x9″x8′6″	Broken Stone		460	920	1,840	3,680	5,520	7,360	9,200
	Gravel		625	1,250	2,500	5,000	7,500	10,000	12,500

Every roadmaster and section-foreman is, or should be, equipped with a tape-line 50 ft. long, divided into feet, inches, halves and quarters of an inch. But there should also be provided for the special work which will fall upon the roadmaster, a steel tape-line which is divided into feet, tenths and hundredths of a foot. All railroad surveyors in this country now use the last-named arrangement, which does away entirely with vulgar fractions, substituting for them the "decimal point." Fig. 246 rep-

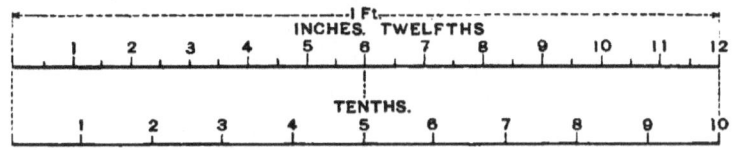

Fig. 246.—Comparison between Inches and Tenths of a Foot.

resents a foot measure which is divided according to these two methods. The upper line contains twelve equal parts which are called "inches." The lower line, although exactly the same length as the upper one, contains but ten equal parts and these are called "tenths." If now a tenth of a foot is divided into ten parts, each of the last-named parts will equal a hundredth of a foot, and again if each hundredth of a foot is divided into ten parts, one of these parts will equal a thousandth of a foot.

The great advantage of the "tenths" is seen when one must add, subtract, multiply or divide several figures. Suppose that it is necessary to add $3\frac{7}{16}$ in., $5\frac{5}{8}$ in., $7\frac{1}{2}$ in. and $9\frac{1}{4}$ in., and get an answer in feet, inches and a fraction. One must first change all these vulgar fractions to sixteenths, then add up the sixteenths, then divide the sum by sixteen and add the result to the inches; then the inches must be added up and divided by twelve to get them into feet. The answer is 2 ft. $1\frac{13}{16}$ in.

But by using the decimal parts of a foot and adding them together like this,

$$3\tfrac{7}{16} = 0.286$$
$$5\tfrac{5}{8} = 0.469$$
$$7\tfrac{1}{2} = 0.625$$
$$9\tfrac{1}{4} = 0.771$$
$$\overline{}$$
$$2.151,$$

the answer is got by one operation and is expressed as two, and one hundred fifty-one thousandths feet. In multiplication and division the use of "tenths" simplifies the operation still more.

Use of Table II.

By means of Table II all of these figures may be got in a moment; look in the column headed 3 inches, opposite $\tfrac{7}{16}$ and .286 will be found; that is, two hundred and eighty-six thousandths of a foot. Or taking another figure in our addition, .625, suppose that it is desired to know how many inches this equals. Look in Table II until .625 is found, when it will be seen that it is under 7 and opposite $\tfrac{1}{2}$, which means seven and one-half inches. If a number must be used at any time which does not exactly agree with the numbers in the table, as for instance .364, then look for the nearest to it which is seen to be .365; this equals $4\tfrac{3}{8}$ in. Thus it is evident that this table may be used either to convert inches to decimals of a foot or decimals of a foot to inches.

Erecting a perpendicular.

How to "erect a perpendicular," in other words lay out a line at right angles from a certain point on another line, is a necessary thing for every trackman to know. In fig. 247, let A–B be the first line, C–D the second line and A the point from which the perpendicular is to be erected. Take a tape and have the end and the 12 ft. mark held together at A; have the 3 ft. mark held on the same

RULES AND TABLES. 233

TABLE II.—INCHES AND FRACTIONS OF AN INCH IN DECIMAL FRACTIONS OF A FOOT.

Parts of an Inch.	0	1	2	3	4	5	6	7	8	9	10	11	Parts of an Inch.
0	.000	.083	.167	.250	.333	.417	.500	.583	.667	.750	.833	.917	0
1/16	.005	.089	.172	.255	.339	.422	.505	.589	.672	.755	.839	.922	1/16
1/8	.010	.094	.177	.260	.344	.427	.510	.594	.677	.760	.844	.927	1/8
3/16	.016	.099	.182	.266	.349	.432	.516	.599	.682	.766	.849	.932	3/16
1/4	.021	.104	.187	.271	.354	.437	.521	.604	.687	.771	.854	.937	1/4
5/16	.026	.109	.193	.276	.359	.443	.526	.609	.693	.776	.859	.943	5/16
3/8	.031	.115	.198	.281	.365	.448	.531	.615	.698	.781	.865	.948	3/8
7/16	.036	.120	.203	.286	.370	.453	.536	.620	.703	.786	.870	.953	7/16
1/2	.042	.125	.208	.292	.375	.458	.542	.625	.708	.792	.875	.958	1/2
9/16	.047	.130	.214	.297	.380	.464	.547	.630	.714	.797	.880	.964	9/16
5/8	.052	.135	.219	.302	.385	.469	.552	.635	.719	.802	.885	.969	5/8
11/16	.057	.141	.224	.307	.391	.474	.557	.641	.724	.807	.891	.974	11/16
3/4	.062	.146	.229	.312	.396	.479	.562	.646	.729	.812	.896	.979	3/4
13/16	.068	.151	.234	.318	.401	.484	.568	.651	.734	.818	.901	.984	13/16
7/8	.073	.156	.240	.323	.406	.490	.573	.656	.740	.823	.906	.990	7/8
15/16	.078	.161	.245	.328	.411	.495	.578	.661	.745	.828	.911	.995	15/16

Erecting a perpendicular. line at D, then when the tape is stretched and brought to an angle at the 8 ft. mark as at B,

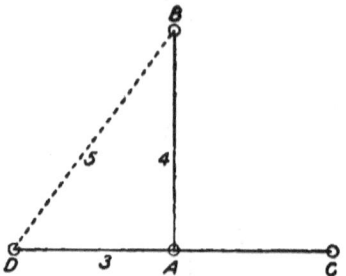

FIG. 247.—Erecting a Perpendicular.

the line joining A and B will be perpendicular to C–D at A. It is seen in fig. 247, that the numbers 3, 4 and 5 are the lengths of the different sides of the triangle A B D. If the figures 3–4–5 are not convenient, any multiple of them may be used instead, if they are all multiplied by the same number as for instance 6–8–10 which are multiples of 2, or 9–12–15 which are multiples of 3.

Letting fall a perpendicular. To "let fall a perpendicular" or in other words lay out a line from a point A, fig. 248,

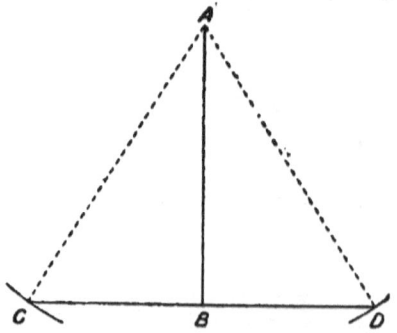

FIG. 248.—Letting Fall a Perpendicular.

at right angles to another line C–D, take a tape or any other cord longer than the distance between A and B and hold one end at A. Have the other end carried first to C and a mark made there, next to D and a mark made there. Then if a mark be placed exactly half way between C and D at B, the line A – B will be perpendicular to C – D from A.

Curves are commonly spoken of with reference to the angle at the center, subtended by a 100 ft. chord. This idea is illustrated in figs. 249, A and B, the first of which repre-

Degree of curve.

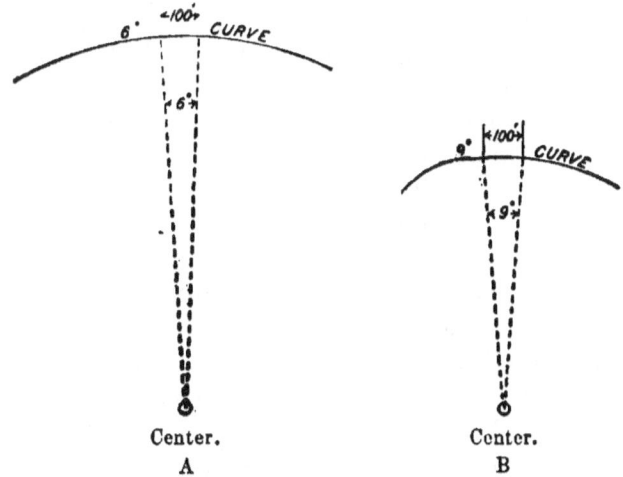

Fig. 249.—A, Six-Degree Curve. B, Nine-Degree Curve.

sents a 6° (six degree) curve and the second a 9° (nine degree) curve.

To find the degree of curve in a railroad track, take a line exactly 62 ft. long and stretch it so that the ends just touch the gage side of the outside rail as in fig. 249; then at

Finding degree of curve.

Fig. 250.—Method of Finding Degree of Curve.

the center of the line (31 ft. from each end) measure from the line to the gage side of the rail and the number of inches found will equal the degree of the curve. That is, if the distance is 3 in., as in fig. 250, it will be a three degree (3°) curve, if $4\frac{1}{2}$ inches a $4\frac{1}{2}°$ curve. By this process the proper elevation for a curve may be found approximately and at any time, with nothing but a measuring tape or a foot rule and ditching-line.

Laying out curves. Long pieces of track which follow a new line and do not run parallel to any old track should, if at all important, be first located with a transit; but short tracks even when on a curve may be quite accurately staked out with a tape line by means of the "versed sine" method or with the assistance of Table III, which gives the deflections for curves up to 20°.

The simplest way of running curves is known as the "versed sine" method; and is illustrated in fig. 251. It has the peculiar advantage of requiring no tables and scarcely any effort of memory, while at the same time it is as correct as any plan can be in which no transit is used.

Assuming that there are two tangents K – A – C and B – C which it is desired to connect by means of a regular curve, the first thing to do is to mark the exact place where the tangents come together at C. This is readily accomplished by setting up two thin stakes on each tangent and sighting them in until C is found to be in both lines. The drawing represents a side-track B – J – E – G – A – K which starts at the frog point B where it is tangent to the frog rail and proceeds on a regular curve to A where it is tangent to the straight line A – K.

After having located C it is necessary to find out the distance to the nearest point from which the curve must start, remembering always that the further A and B are from C, the easier the curve will be. If as in the drawing the curve must begin not further away than the frog point, that fact at once limits the distance B – C but if the curve is removed from any other track and is to join two simple tangents then there is no reason why it should not begin anywhere else back of A or B as at K.

Laying out curves.

When the distance C – B has been determined, next mark the distance C – A exactly the same as C – B; lay off the straight line A – B and place a stake at A exactly half way between and exactly in line with A – B. Then measure the distance C – D; exactly half way between and exactly in line with C and D place the stake E. This stake will be on the curve. Next join A – E and half way between place the stake F; join E – B and half way between place H, then on a line perpendicular

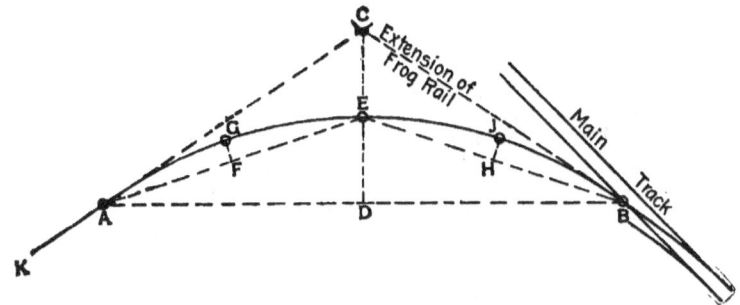

FIG. 251.—Versed Sine Method of Laying Out Curves.

to A – E at F and on a line perpendicular to E – B at H lay off F – G and H – J, each of them exactly one quarter of the distance from E to D. The stakes at B – J – E – G and A are then all of them in the line of a curve which is tangent to straight lines at A and B.

If the curve is not more than two hundred feet long, these five stakes are enough to locate it but if for any reason more points are needed,

Laying out curves they may easily be supplied by joining A and G, G and E, E and J, J and B by straight lines; then exactly half way on these lines and perpendicular to them, lay off other points one quarter of the distance between G and F.

Sometimes an obstruction or the character of the ground will interfere with producing both of these tangents as far as C, in which case a recourse must be had to another plan which is equally correct but, because it involves the use of Table III and requires the location of more points, is not quite so convenient.

Like the preceding plan, the one now to be described will usually be needed for laying out new side tracks and, because of this, the curve is shown, in fig. 252, as beginning at a frog.

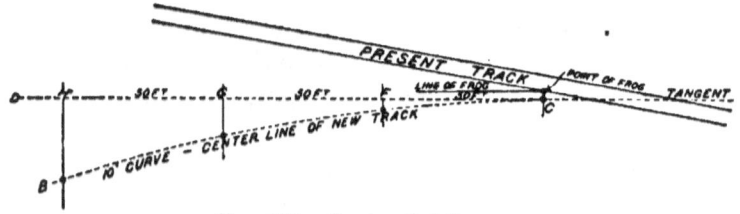

FIG. 252.—Laying Out Curves.

There is no reason however, if the circumstances are different, why the curve should not begin at any other point, in which case the 50-ft. spaces may be laid off and the curve located in exactly the same way, except that there will be no frog to start from.

In fig. 252, the line C D represents the "tangent" (the straight line from which the curve starts) which is parallel with the outside gage line of the frog, and 2 ft. $4\frac{1}{4}$ in. (or 2 ft. $4\frac{1}{2}$ in. with a 4 ft. 9 in. gage) distant from it. This is the most correct way but if it be preferred, the "line of frog" may be used as the tangent with results which are practically just as good. In consequence of this offset C is on the center line of the new track at the frog.

Laying out curves.

Since it is best to start the curve at the heel of the frog, C is located opposite that place and becomes the point of curve. C and D are marked on the ground by slender sticks about 4 ft. long, set upright. Beginning at C, in line with these two sticks, lay off the 50, 100, 150 ft., etc. marks and then at these last named points erect the lines F, G, H perpendicular to the line C–D, according to the rule given in connection with fig. 247; mark these lines with stakes at F, G and H and also beyond where the curve is likely to reach. On F, G, H lay off the distances given in Table III for the curve that has been decided upon. If it is to be a 10 deg. curve the radius will be 574 ft. long and the respective distances on the lines F, G, H in fig. 249, will be 2.2 ft., 8.8 ft. and 20.0 ft. This method is sufficiently accurate for all ordinary purposes if care is used in locating the different points.

In order to find out what curve is required from a certain tangent in order to strike a certain point in a railroad track, the same method already described may be used or else the reverse of it which is as follows. Suppose for instance that in front of a factory, as in fig. 253, from a tangent A–C it is desired to lay off a curve which will not reach beyond C and will be tangent to the line of frog near E.

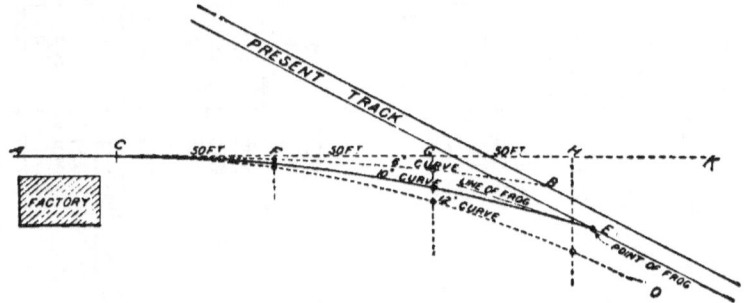

Fig. 253.—Laying Out Curves.

TABLE III.—Curves with their Radii and Offsets from a Tangent.

Radius.	Degree.	\multicolumn{9}{c}{Distances on Tangent, Feet.}
		50	100	150	200	250	300	350	400	500
		\multicolumn{9}{c}{Offsets — Feet and Tenths.}
5,730	1°	0.2	0.9	2.0	3.5	5.4	7.8	10.7	14.0	21.9
2,865	2°	0.4	1.7	3.9	7.0	11.0	15.8	21.5	28.1	44.0
1,910	3°	0.7	2.6	5.9	10.5	16.5	23.8	32.3	42.3	66.6
1,433	4°	0.9	3.5	7.9	14.1	22.0	31.8	43.4	57.0	90.5
1,146	5°	1.1	4.4	9.9	17.6	27.6	40.0	54.7	72.2	114.9
955	6°	1.3	5.2	11.9	21.2	33.3	48.4	66.4	87.8	
819	7°	1.5	6.1	13.9	24.8	39.1	56.9	79.5	104.4	
717	8°	1.8	7.0	15.9	28.5	45.0	65.8	91.2		
637	9°	2.0	7.9	18.0	32.2	51.1	75.1	104.8		
574	10°	2.2	8.8	20.0	36.0	57.3	84.7			
522	11°	2.4	9.7	22.1	39.9	63.8	94.4			
478	12°	2.6	10.6	24.1	43.9	70.6	105.9			
442	13°	2.9	11.5	26.2	47.9	77.5				
410	14°	3.1	12.4	28.4	52.1	85.0				
383	15°	3.3	13.3	30.6	56.4	92.8				
359	16°	3.5	14.2	32.8	60.9	101.3				
320	18°	4.0	16.0	37.3	70.2					
288	20°	4.5	18.0	42.1	80.6					

Laying out curves. First stake off the tangent A − C − K parallel with the factory and locate the 50 ft. points beginning at C. Erect the perpendiculars at F, G, H and then by means of Table III, lay off any curve, as C − B, which is a 6° curve. This is too flat so next a 12° curve, C − D is tried; this proves too sharp but after continual trials the right one will be found which in fig. 253 is seen to be C − E, a 10° curve.

Frequently in this method, it will be found that the curve, although it strikes near E, will not fit the frog. In this case the point C must be moved nearer to E and new curves tried until the right one is found. But no matter what trouble is experienced at first, let it be remembered that any trackman can use this table successfully if he will only try.

Curving rails. In bending rails to fit the curves at switches and in the main track, Table IV will be found useful.

Fig. 254.—Bending Rails.

Fig. 254 represents a 26 ft. rail which is to be made to conform to an 11° curve. Take a line and stretch it on the gage side of the head from one end of the rail to the other; mark the middle at A, 13 ft. from each end, and the quarters B and C, 6½ ft. each from the middle and the ends. According to Table IV, when the rail is properly bent, the perpendicular distance A (middle ordinate) from the line to the rail-head will be 2 inches, while B and C (ordinates at the quarters) will each of them be ¾ of this distance, which is 1½ inches.

Frog numbers and angles. Frogs which are used with a switch, are most often described by their numbers, but sometimes according to the angle formed by

TABLE IV.—MIDDLE-ORDINATES FOR CURVING RAILS.

(Ordinates at the quarters are 3/4 of Middle-Ordinates.)

DEGREE OF CURVE.	LENGTH OF RAILS (Feet).											DEGREE OF CURVE.
	30	28	26	24	22	20	18	16	14	12	10	
	INCHES.											
1°	1/4	3/16	3/16	1/8	1/8	1/8	1/16	1/16	1/16	1/16	1/16	1°
2°	1/2	7/16	3/8	5/16	1/4	1/4	3/16	1/8	1/8	1/16	1/16	2°
3°	11/16	5/8	9/16	7/16	3/8	5/16	1/4	1/4	3/16	1/8	1/16	3°
4°	15/16	7/8	3/4	5/8	1/2	1/2	3/8	5/16	1/4	3/16	1/8	4°
5°	1 3/16	1 1/16	7/8	3/4	5/8	9/16	7/16	3/8	1/4	3/16	1/8	5°
6°	1 7/16	1 1/4	1 1/16	15/16	13/16	5/8	1/2	7/16	5/16	1/4	3/16	6°
7°	1 11/16	1 1/2	1 1/4	1 1/16	7/8	3/4	5/8	1/2	3/8	1/4	3/16	7°
8°	1 15/16	1 11/16	1 7/16	1 3/16	1 1/16	7/8	11/16	9/16	1/2	5/16	1/4	8°
9°	2 1/8	1 7/8	1 5/8	1 3/8	1 1/8	15/16	3/4	5/8	1/2	3/8	1/4	9°
10°	2 3/8	2 1/16	1 13/16	1 1/2	1 5/16	1 1/16	7/8	11/16	9/16	3/8	1/4	10°
11°	2 5/8	2 1/4	2 –	1 11/16	1 7/16	1 3/16	15/16	3/4	5/8	7/16	5/16	11°
12°	2 7/8	2 1/2	2 3/16	1 13/16	1 9/16	1 1/4	1 1/16	7/8	5/8	1/2	5/16	12°
13°	3 1/16	2 11/16	2 5/16	2 –	1 11/16	1 3/8	1 1/8	15/16	11/16	1/2	3/8	13°
14°	3 5/16	2 7/8	2 1/2	2 1/8	1 13/16	1 1/2	1 3/16	1 –	3/4	9/16	3/8	14°
15°	3 9/16	3 1/8	2 11/16	2 1/4	1 15/16	1 9/16	1 5/16	1 1/16	13/16	5/8	7/16	15°
16°	3 3/4	3 5/16	2 7/8	2 7/16	2 1/16	1 11/16	1 3/8	1 1/8	7/8	5/8	7/16	16°
17°	4 –	3 1/2	3 1/16	2 9/16	2 3/16	1 13/16	1 7/16	1 3/16	7/8	11/16	7/16	17°
18°	4 1/4	3 11/16	3 3/16	2 11/16	2 5/16	1 7/8	1 9/16	1 1/4	15/16	11/16	1/2	18°
19°	4 1/2	3 7/8	3 3/8	2 7/8	2 7/16	2 –	1 5/8	1 5/16	1 –	3/4	1/2	19°
20°	4 3/4	4 1/8	3 9/16	3 –	2 9/16	2 1/8	1 11/16	1 3/8	1 1/16	13/16	9/16	20°
21°	4 15/16	4 5/16	3 3/4	3 3/16	2 11/16	2 3/16	1 13/16	1 7/16	1 1/8	7/8	9/16	21°
22°	5 3/16	4 1/2	3 15/16	3 5/16	2 13/16	2 5/16	1 7/8	1 1/2	1 3/16	7/8	9/16	22°
23°	5 7/16	4 11/16	4 1/16	3 7/16	2 15/16	2 3/8	1 15/16	1 9/16	1 3/16	15/16	5/8	23°
24°	5 5/8	4 15/16	4 1/4	3 5/8	3 1/16	2 1/2	2 1/16	1 11/16	1 1/4	15/16	5/8	24°
25°	5 7/8	5 1/8	4 7/16	3 3/4	3 3/16	2 5/8	2 1/8	1 3/4	1 5/16	1 –	11/16	25°
26°	6 1/16	5 5/16	4 5/8	3 7/8	3 5/16	2 11/16	2 3/16	1 13/16	1 3/8	1 –	11/16	26°
27°	6 5/16	5 1/2	4 3/4	4 1/16	3 7/16	2 13/16	2 5/16	1 7/8	1 7/16	1 1/16	11/16	27°
28°	6 9/16	5 11/16	4 15/16	4 3/16	3 9/16	2 15/16	2 3/8	1 15/16	1 7/16	1 1/8	3/4	28°
29°	6 13/16	5 7/8	5 1/8	4 3/8	3 5/8	3 –	2 7/16	2 –	1 1/2	1 1/8	3/4	29°

RULES AND TABLES.   243

the two running rails; this latter method is always used when speaking of crossing frogs. **Frog numbers and angles.**

The number of a frog is determined by dividing distance, B, in fig. 255, into distance, A,

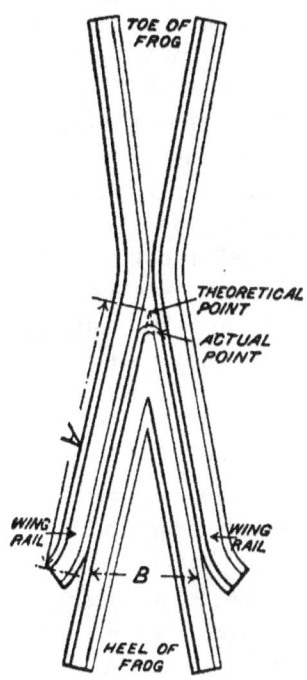

FIG. 255.—Frog Numbers and Angles.

remembering always to use the true (or "theoretical") point of the frog; this is not the end of the rails but may be found by laying a straight edge on each gage side of the frog and marking the place where the straight edges come together. As an example let it be supposed that A equals 48 inches, and B equals 8 inches. Then 48 divided by 8 equals 6, and the frog is therefore a number 6.

Since the object of this calculation is only to find the proportion existing between the length and the width of the frog, a divided measure is not at all necessary for it; anything, a lead pencil or a stick, which is shorter than

Frog numbers and angles. the width of the frog at the heel, will do. Place the article where its length is exactly equal to the distance between the gage lines and measure with it from there to the true point. The number of lengths made in the last measurement equals the number of the frog; that is if the place from which the measurement started is six times as far from the point as the lead pencil is long, it is a number six frog.

When only the angle of a frog is known and the number is also desired, first reduce the angle to minutes (there are sixty minutes in a degree) and then divide 3440 by it. The result will equal the number of the frog.

Example: What is the number of a 5° 44′ (five degrees and forty-four minutes) frog?

5° 44′ = 5 × 60 + 44 = 344′. 3440 divided by 344 = 10. That is, a No. 10 frog.

Conversely, when only the number is known and the angle is also desired divide 3440 by the number of the frog and reduce the result to degrees and minutes.

Example: What is the angle of a No. 8 frog?

3440 divided by 8 = 430′, and 430 divided by 60 = $7\frac{10}{60}$ = 7° 10′. That is a seven degree and ten minutes (7° 10′) frog.

Switch leads. A common rule for the calculation of a switch lead was to multiply twice the gage of the track by the number of the frog. This rule is well enough for the shorter leads, but in the case of a No. 10 frog, it amounts to a distance of 94 feet, which besides being unnecessarily great, requires that the point of the switch rail shall be planed to a too fine point. The method of calculating has therefore been changed in the table contained in this volume. It will be noticed in Table V that an 18 ft. point is provided for the 11 and 12 split switch leads. This was done because those two frogs are seldom used except where

Switch leads.

trains are expected to run fast; in that case, the easier the bend at the main track, the better. With the 4 and 5 split switches a 10 ft. point is arranged for, because these frogs should not be used except to make the lead as short as possible; hence the necessity of contributing to this object in every legitimate way. In determining the leads and cross distances of all the switches, a regular curve is assumed to begin at the heel of the switch and continue to within exactly 5 ft. of the theoretical point of frog. It is believed that with the distances shown in fig. 256 and the corresponding amounts in Table V, any of the switches named there, may be put in accurately and without difficulty. The distances A, B, C and D are all of them to be marked with chalk on the main track rail, measuring from the theoretical point of frog as a starting point. Then at these places and from the gage side of the main track rail to the gage side of the side track rail, the distances a, b and c are to be laid off perpendicular to the main rail. The distance d of course, is nothing but the gage minus the offset of 6 in. for split switches, and the gage minus the 5 in. throw for stub switches, since these are constant distances for all numbers of frogs. For convenience, the diagrams (fig. 256) are shown with a straight main track; but it is to be understood that if the main track is curved, the degrees of curves and the radii of the side track rail will be different from those named in Table V. If the frog is in the outer rail of a curved main track, the degree of curve of the lead will be equal to the degree of curve given in the table, *minus* the degree of curve of the main track. If the frog is in the inner rail of a curved main track, the degree of curve of the lead will equal the degree of curve

THE NEW ROADMASTER'S ASSISTANT.

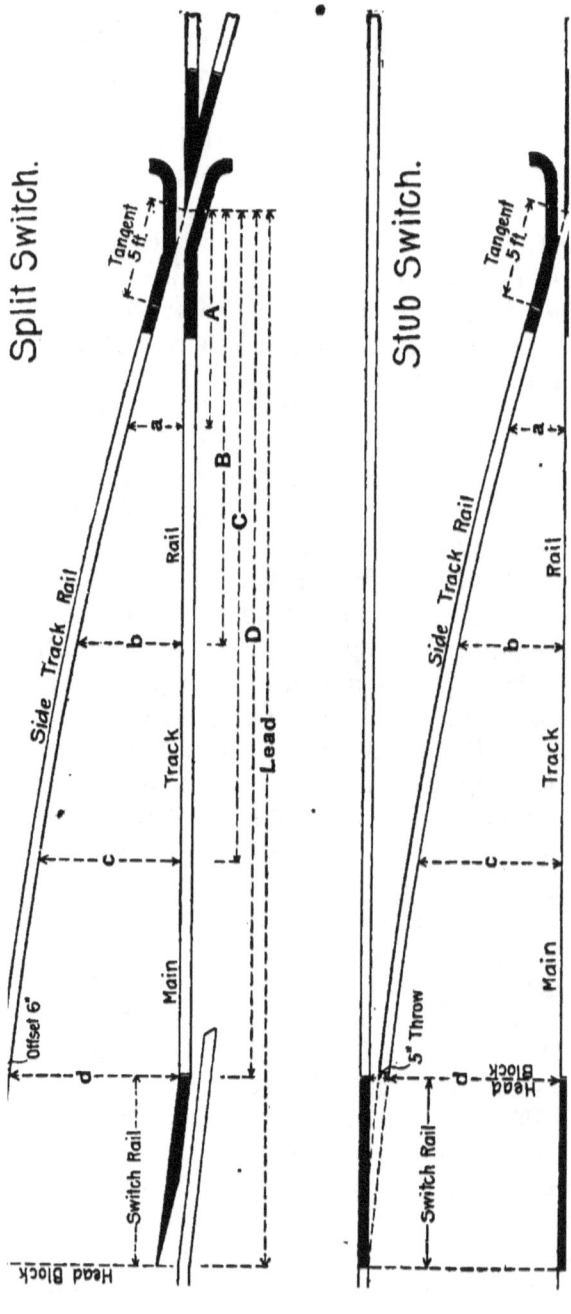

TABLE V.—LEADS FOR SPLIT AND LEADS FOR STUB SWITCHES—4 ft. 8½ in. Gage.

Offset at Heel of Split Rail = 6 in. Throw of Stub Switch = 5 in. Add ¼ in. to "c" and ½ in. to "d" for 4 ft. 9 in. gage.

			4	5	6	7	8	9	10	11	12
FROGS:	Number		4	5	6	7	8	9	10	11	12
	Angle		11° 22′	11° 30′	9° 34′	8° 12′	7° 10′	6° 22′	5° 44′	5° 12′	4° 47′
OUTER RAIL OF SWITCH:	Degree of Curve		60° 40′	34° 00′	22° 48′	15° 56′	11° 40′	8° 48′	6° 53′	5° 42′	4° 42′
	Radius		99 Ft.	171 Ft.	253 Ft.	361 Ft.	494 Ft.	651 Ft.	833 Ft.	1,005 Ft.	1,222 Ft.
LENGTH OF SWITCH RAIL:	Split Switches		10 Ft.	10 Ft.	15 Ft.	15 Ft.	15 Ft.	15 Ft.	15 Ft.	18 Ft.	18 Ft.
			Ft. In.	Ft. In.	Ft. In.	Ft. In.	Ft. In.	Ft. In.	Ft. In.	Ft. In.	Ft. In.
	Stub Switches		10-11	12-9	16-8	18-6	20-8	23-4	25-0	25-0	25-0
DISTANCES ALONG MAIN RAIL:	Alike for Split and Stub Switches,	A	6-1¼	7-7¼	9-7¾	11-1½	12-6½	13-10½	15-1¾	17-1	18-3½
		B	12-2½	15-2½	19-3½	22-3	25-1	27-9¼	30-3¼	34-1¾	36-6¾
		C	18-3¾	22-9½	28-11	33-4½	37-7½	41-7¾	45-5¼	51-2¾	54-10¼
	Split Switches,	D	24-5¼	30-4¾	38-6¾	44-6	50-2	55-6¼	60-7	68-3½	73-1¾
	Stub Switches,	D	26-1¼	32-0¾	41-0¾	47-0	52-8	58-0¼	63-1	71-3½	76-1¾
LENGTH OF LEAD:	Split Switches		34-5¼	40-4¾	53-6¾	59-6	65-2	70-6¼	75-7	86-3½	91-1¾
	Stub Switches		26-1¼	32-0¾	41-0¾	47-0	52-8	58-0¼	63-1	71-3½	76-1¾
DISTANCES ACROSS:	Alike for Split and Stub Switches,	a	1-7	1-6½	1-7	1-6¾	1-6¼	1-6	1-5½	1-5¾	1-5½
		b	2-10	2-9¼	2-10	2-9½	2-8	2-8¼	2-8	2-8¼	2-7¾
		c	3-8½	3-8	3-8¼	3-8	3-7½	3-7¼	3-6¾	3-7	3-6¾
	Split Switches,	d	4-2½	4-2½	4-2½	4-2½	4-2½	4-2½	4-2½	4-2½	4-2½
	Stub Switches.	d	4-3½	4-3½	4-3½	4-3½	4-3½	4-3½	4-3½	4-3½	4-3½

given in Table V *plus* the degree of curve of the main track. The distances along the main track rail and the cross distances will remain, for all practical purposes, the same as though the main track were straight.

Three-throw switches.

The combinations which are possible between two switches are so numerous that no table can be framed which will begin to meet the requirements. With split switches these combinations are infinite as is evident from fig. 257, where are illustrated the three bases of procedure. In A, which is comparatively rare, the points begin at practically the same place which results in a lead to all intents the same as that of a three-throw stub switch. This is the plan followed in Table VII. B assumes that the frogs are placed opposite each other and that the switch-point of the short lead is far enough back of the other switch-point to leave room for the "throw" and the attachment of the connecting-rod. This is not a very rational method since it leaves too small a choice of frogs to meet the special cases which will surely arise. Plan C is the one which offers the greatest variety. Since a three-throw switch should never be used if there is any way of putting in two entirely separate single switches this plan is the best one of the three that have been described. It is not only the best because it permits the greatest variety of leads but because it more nearly secures the same track conditions that exist when two entirely separate leads are used.* Although, as has been stated, it is impossible to frame a table fulfilling these conditions, fig. 258 and Table VII present one

---

* In that useful work already mentioned, "Switch Layouts and Curve Easements," Mr. Torrey has included nearly a hundred diagrams of various three-throw switch leads which contain all the information necessary for putting them in, together with a table of switch-timbers for each one. Table IV is adapted from

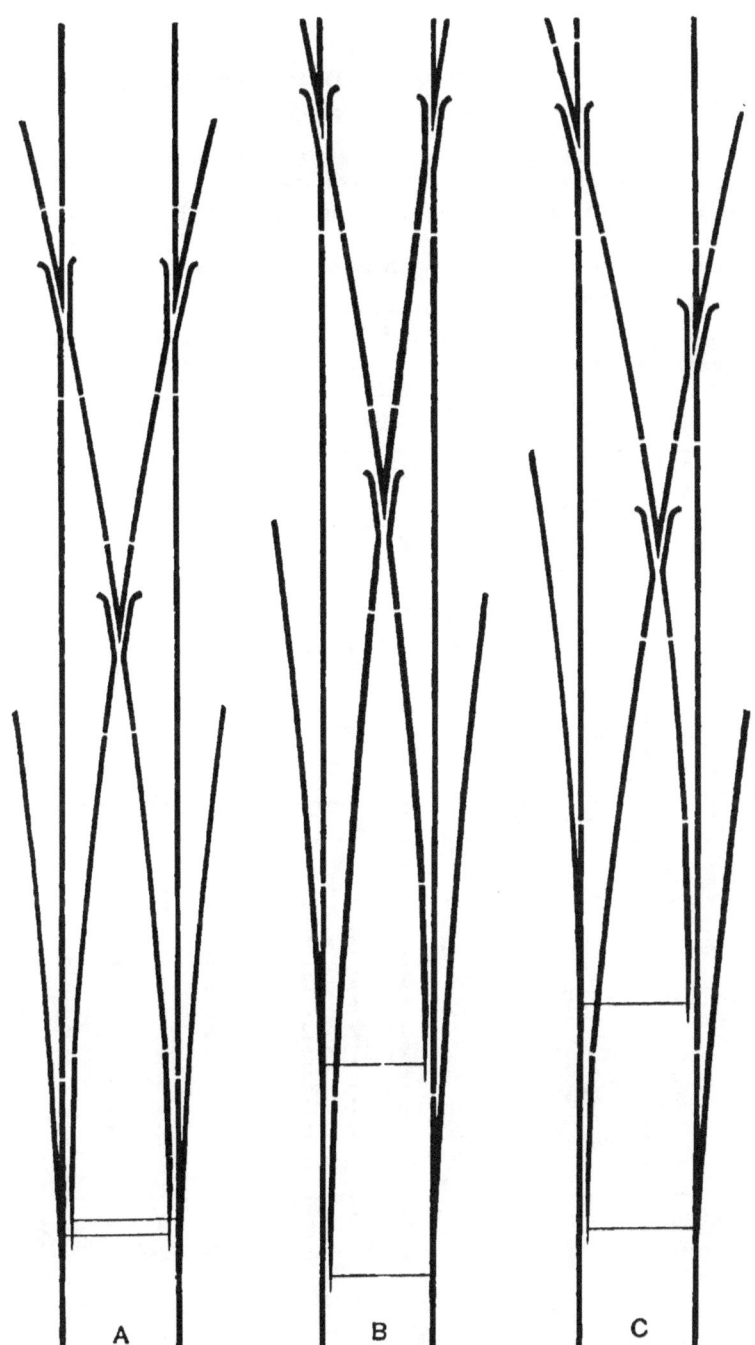

Fig. 257.—Methods of Arranging Three-Throw Split Switches.

TABLE VI.—NUMBER OF SWITCH-TIES FOR USE WITH TABLES V AND VII.

1—Head-Block 8 in. by 12 in.—16 ft. long with each Single Split and Stub Switch and each Three-Throw Stub Switch.
2—Head-Blocks 8 in. by 12 in.—16 ft. long with each Three-Throw Split Switch.
All other Switch-Ties 7 in. by 9 in.—20 in. center to center.
Standard Cross-Ties 8 ft. 6 in. long.

Number of Frog				4	5	6	7	8	9	10	11	12
			Feet Long	\multicolumn{9}{l}{Number of Pieces 7 in. by 9 in.}								
Single Switches	Stub Switch,		9	1	1	3	3	3	3	3	2	2
	Split Switch,			6	6	10	10	10	10	10	11	11
	Alike for Split and Stub Switches		10	6	7	8	9	10	10	11	12	13
			11	3	4	6	6	7	9	9	11	11
			12	2	3	4	5	6	6	7	9	9
			13	2	3	3	4	5	6	7	7	8
			14	3	3	3	5	4	5	6	6	7
			15	2	3	4	4	5	5	6	6	8
Three-Throw Switches	Stub Switch,		9	—	—	—	—	—	—	—	—	—
	Split Switch,			3	3	5	5	5	5	5	5	5
	Stub Switch,		10	3	3	5	5	6	6	6	5	6
	Split Switch,			5	5	7	7	8	8	8	9	10
	Alike for Split and Stub Switches		11	2	3	5	5	5	5	6	7	8
			12	2	3	3	4	4	5	5	6	6
			13	2	2	3	3	4	4	4	5	5
			14	1	2	2	3	3	4	4	5	5
			15	1	1	2	2	3	3	4	4	4
			16	1	2	2	2	2	3	4	4	4
			17	2	1	1	2	2	3	3	4	4
			18	1	2	2	2	3	3	3	3	4
			19	1	1	2	2	2	3	3	3	4
			20	1	2	2	2	3	2	3	3	3
			21	2	1	1	2	2	3	3	3	3
			22	1	2	2	2	2	3	3	3	4
			Feet Long									
Number of Frog.				4	5	6	7	8	9	10	11	12

SPLIT SWITCHES.

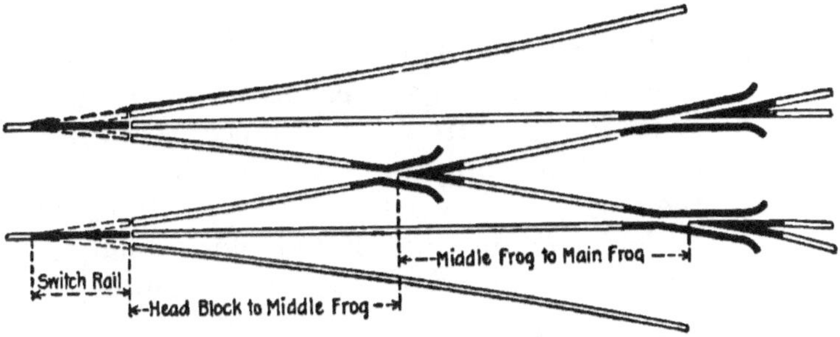

STUB SWITCHES.

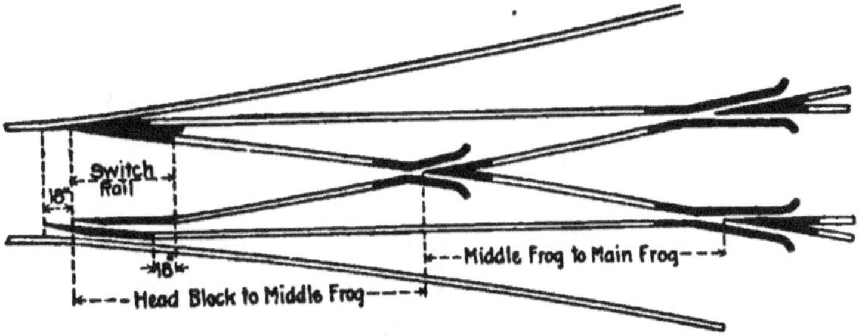

FIG. 258.—Diagram to accompany Table VII.

TABLE VII.—THREE-THROW SPLIT AND STUB-SWITCH LEADS.

Offset at heel of Split Switches = 6 inches.

Throw of Stub Switches = 5 inches in each direction.

Left-hand Split Switch to be placed 18 inches in advance of Right-hand Split Switch.

Number of Main Frogs.	Middle Frog Angle.		Middle Frog to Main Frog.		Head Block to Middle Frog.				Switch Rail Length.			
					Split Switch		Stub Switch		Split Switch		Stub Switch	
	Deg.	Min.	Feet	Inches	Feet	Inches	Feet	Inches	Feet	Inches	Feet	Inches
4	22	58	9	8½	24	9	16	6	10	0	10	11
5	17	52	12	5¼	28	0	19	5½	10	0	12	9
6	14	30	15	2¾	38	4	25	8½	15	0	16	8
7	12	14	18	0¼	41	7	29	1	15	0	18	6
8	10	38	20	9½	44	4½	31	10¾	15	0	20	8
9	9	28	23	6	47	0	34	6½	15	0	23	4
10	8	32	26	2	49	5	37	0	15	0	25	0
11	7	40	29	0½	57	3	42	1½	18	0	25	0
12	7	04	31	7¼	59	6	44	8	18	0	25	0

Locating frogs.

method of putting in both split and stub 3-throw-switches.

The ordinary rule for determining the distance between frogs in crossovers is to subtract twice the gage from the distance between centers and multiply the result by the number of the frog. This does perfectly well for frogs of large number but is not close enough for number 4 and 5 frogs. As for example—

```
Distance between centers =   -   - 13.00 ft.
Gage = 4.71 ft.   Twice 4.71 =  -    9.42 ft.
                                     3.58
Number of frog =   -   -   -   -   -    4
                                 -  14.32 ft.
```

That is, 14 ft. 4 in. from frog to frog along the main rail.

But Table VIII, under the heading of 13 ft. between centers of tracks, gives this distance

TABLE VIII.—Distance D. Frog to Frog along the Main Rail in Cross-overs—Gage, 4 ft. 8½ in.

Frog Numbers	Distance Between Centers of Tracks.						
	Ft. In. 11-0	Ft. In. 11-6	Ft. In. 12-0	Ft. In. 12-6	Ft. In. 13-0	Ft. In. 13-6	Ft. In. 14-0
4	5-7	7-6	9-6	11-5	13-5	15-4	17-4
5	7-4	9-9	12-3	14-8	17-2	19-7	22-0
6	9-0	12-0	14-11	17-11	20-10	23-10	26-9
7	10-8	14-1	17-7	21-1	24-6	28-0	31-5
8	12-4	16-3	20-3	24-3	28-3	32-2	36-2
9	13-11	18-5	22-11	27-5	31-10	36-4	40-10
10	15-6	20-6	25-6	30-6	35-6	40-5	45-5
11	17-2	22-8	28-2	33-8	39-2	44-8	50-2
12	18-9	24-9	30-8	36-8	42-8	48-7	54-7

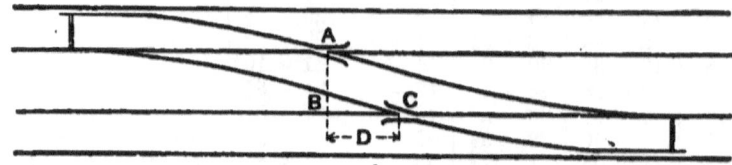

Fig. 259.—Diagram to accompany Table VIII.

as 13 ft. 5 in., a difference of nearly a foot. In placing the frogs according to the above rule, or Table VIII, first let fall a perpendicular from one of the frog points, A fig. 259, to the nearest rail of the other track at B. From B lay off the proper distance D; then C will be the location of the other frog point.

INDEX.

A.

Alexander car replacer, 196.
American nut-lock, 111.
American Steel Foundry, ditching machine, 58.
Anchor fence post, 22
Angle bars (see rail joints).
Avery steel fence post, 23.

B.

Ballast, 75.
  stone, 76.
  conveyers, 77.
  crushers, 78.
  cleaning, 79.
  gravel, 79.
  plow, 82.
  unloader, 83.
  cars, 84.
  heaving, 86.
  sections, 87–89, 91.
  drains, 90.
  picks, 156.
  forks, 158.
  car capacity for, 229.
  quantities per mile, 230.
Barnhart ballast plow, 82.
Barrett track jack, 92.
Bell-crank, 219.
Bending (see curving rails).
Block signaling, 224.
  banjo signal, 225.
  banner signal, 226.
  track circuit, 226.
Bolt-lock, 216.
Boyer & Radford track jack, 93.
Bogue & Mills crossing gate, 44.
Bond steel fence post, 23.
Bracket post, 208.
Bridge floor, 69.
  ballasted, 70.
  protected, 71.
  shimming, 72.
Bridge warning, 40.
Bryant rail saw, 146.
Bucyrus steam shovel, 81.
Buda Foundry Co., drills, 144.
  hand car, 151.

Bumper, curved rail, 32.
  clamped, 32.
  triangular, 32.
  timber, 33.
  braced spring, 33.
  Ellis, 33.
Bush cattle guard, 25.
Bush interlocking bolt, 67.

C.

Car replacers, 196.
Cattle guards, 24.
  Bush, 25.
  Kalamazoo, 25.
  National, 25.
  Standard, 26.
  Merrill-Stevens, 26.
Caution signs, use of, 18.
Chain wheel, 220.
Churchill joint, 117.
Clamp for tape line, 155.
Claw bar, 160.
Compensators, 220.
Competition, 6.
Continuous joint, 116.
Creeping rails, 137.
Crossings, signs, 36.
  bells, 41.
  gate (pneumatic), 43.
  continuous, 179.
  narrow angle, 180.
  Fontaine, 181.
  wide-angle, 181.
  steam and street, 181.
Cross ties (see ties).
Culverts, cleaning, 2.
  pipe, 68.
  wooden, 69.
Curves, elevation on, 134.
  elevation on bridges, 135.
  easement, 136.
  widening gage on, 136.
  degree of, 235.
  laying out, 236–240.
  radii and offsets, 240.
Curving rails, 140.
  ordinates for, 241.

D.

Decimals of a foot in inches, 231–233.
Detector bars, 214.
Discipline, 5.
Distant signal, 205.
Ditches, cleaning, 2.
  paving, 58.
  sodding banks, 60.
  slope, 61.
  slope gage, 62.
Ditching, methods, 57.
  machine, 57.
  shovel, 161.
Drainage, 55.
Dump-car (Goodwin), 84, 85.
Dwarf signal, 207.

E.

Elliot Frog & Switch Co., rail brace, 137.
  spring-rail frog, 187.
Ellis bumping post, 33.
Embankments, 63.
Emergencies, 193.
Emerson rail-bender, 141.
Erie railroad ballast sections, 87.
Eureka nut-lock, 111.
Eureka spring-rail frog, 187.
Eyeless tools, 157.

F.

Facing-point lock, 213.
Fairbanks, Morse & Co., rail-bender, 141.
  one-man velocipede, 148.
  gasoline motor, 150.
  push car, 152.
  foot guard, 177.
Feet and inches (decimals), 231–233.
Fence, 19 (see wire fences).
  gangs, 24.
Fence-posts, (see posts).
Filling blocks, 138.
Fires, 17.
Fisher joint, 115.
  offset splice, 131.
Fixed signals, 201.
Flag holder, 163.
Fontaine crossing, 181.
Foot guards, 176.

Foremen, as laborers, 13.
  residence, 14.
Frogs, inspection, 2.
  angles and numbers, 165, 242–244.
  movable, 179.
  rigid plate, 183.
  rigid yoke, 183.
  rigid bolted, 184.
  spring rail, 185–187.
  putting in, 252, 253.

G.

Gangs, combining of, 12.
Gasoline motors, 149.
Gate, farm, 21.
Goodwin dump car, 84, 85.
Grass, 3.
Gravel, ballast, 79.
  pits, 79.
  distributing, 80.
Grip nut, 113.
Guard rails, 174, 175.
  fastener, 176.

H.

Haarmann-Vietor rail, 122.
Hammers, 158.
Hand-cars, use and abuse, 4.
  types of, 151.
Harp switch-stand, 191.
Hartley & Teeter velocipede, 149.
Harvey nut lock, 112.
Hawks offset splice, 131.
Highway crossing, 19.
  open, 27.
  old-rail protection, 27.
  bells, 41.
  gates, 43.
Hollow tires, 17.
Home signals, 204.
Hydraulic ram, 48.

I.

Inches, decimals, 231–233.
Interlocking signals, 201.
  machine, 203.
  levers, 221.
  pneumatic, 223.
Intoxicants, 6.

J.

Jack-knife switch-stand, 191.
Jaws (interlocking), 219.

Jenne track jack, 92.
Jim crow, 140.
Joints (see rail joints).

K.

Kalamazoo cattle guard, 25.
    one-man velocipede, 147.
    safety velocipede, 148.
    gasoline motor, 150.
    early hand-car, 151.
Katte rail, 124.
Knots, 194.

L.

Lamp signals, 200.
Lazy jack, 220.
Leads for switches, 244–251.
Lee's ballasted trestle, 70.
Lidgerwood unloader, 83.
Lining bar, 160.
Long truss joint, 118.

M.

Mail cranes, 34.
Material, location of, 6.
    extra, 193.
    quantities per mile, 230.
Mattock, 156.
Merrill-Stevens cattle guard, 26.
Metal posts, 22, 38, 39.
Mile posts, 37.
Monument, 39.

N.

Napping hammer, 158.
National cattle guard, 25.
    nut lock, 112.
New road, finishing, 1.
New York Central ballast sections, 89.
Nuts, quantities per mile, 230.
Nut locks, 111.
    quantities per mile, 230.

O.

Offset splices, 131.
Oliver Iron & Steel Co., grip nut, 113.
O'Neil crossing bell, 42.
    track instrument, 42.
Opposite and broken joints (see rail joints).

P.

Paint, 41.
Pennsylvania Railroad ballast sections, 88.
Pennsylvania Steel Co., 3-throw switch, 170.
    automatic switch stand, 189.
Perpendiculars, erecting and letting fall, 234.
Picks, 156.
Pinch bar, 160.
Pipe and wire adjusters, 220.
    carriers, 218.
Platforms, 19.
    terra-cotta, 28.
Pneumatic interlocking, 223.
Poage water crane, 51.
Pole drains, 60.
Posts, 19.
    anchor, 22.
    Bond steel, 23, 38, 39.
    Avery, 23.
    distances apart, 24.
    mile, 37.
Post-hole shovels, 162.
Promotion, 7.
Pumps, 48.

Q.

Q & C Co., Servis tie-plate, 98.
    Bryant rail saw, 146.

R.

Rail, form and comparisons, 107.
    ends, 121.
    Haarman-Vietor, 122.
    Katté, 124.
    long, 124.
    continuous, 124.
    welded, 125.
    counting and turning, 127.
    unloading, 128.
    re-laying, 129.
    spacing, 130.
    time to relay, 132.
    short pieces, 132.
    braces, 137.
    expansion device, 139.
    curving, 140.
    benders, 140.
    punch, 142.
    drilling, 143.
    drills, 143.

INDEX.

Rail—*(Continued)*.
  cutting, 145.
  saws, 146.
  tongs, 162.
  fork, 163.
  at switches, 174.
  quantities per mile, 230.
Rail joints, 113–118.
  suspended and supported, 119.
  offset, 131.
  opposite and broken, 133.
Rainy days, 12.
Ramapo automatic switch stand, 188.
Ratchet drills, 143.
Re-ballasting, 86.
Re-laying rails, 129.
Reports, 4.
Retaining walls, 63.
  foundations of, 65.
Re-railing device, 71.
Reverse pointed spike, 67.
Road (highway), 28.
Roadmaster, duties, 7.
Road bed, section, 55.
Roberts, Throp & Co., velocipede, 149.
  hand car, 152.
  push car, 152.
  foot guard, 177.
Rodger ballast car, 84.
  plow car, 84.
Routine work, 12.

S.

Samson joint, 114.
Section, house, 31.
  men, number per mile, 9.
  length of, 10.
Selector, 216.
Separation of grades, 45.
Servis tie plate, 98.
Sheffield water crane, 52.
  one-man velocipede, 148.
  gasoline motor, 150.
  push car, 152.
  foot guard, 177.
Shimming tool for bridges, 130.
Shims, 104.
Shovels, 161.
Sign, crossing, 36.
  bridge, etc., 37.

Sign—*(Continued)*.
  letters, 40.
Signals, train, 197.
  whistle, 199.
  lamp, 200.
  fixed, 201.
  interlocking, 201.
  home, 204.
  distant, 205.
  dwarf, 207.
  block, 224.
Six-hole angle-bar, 114.
Sledges, 159.
Smith rail saw, 146.
Snow shovel, 161.
  storms, 3.
Spike, reverse pointed, 67.
  holes, 103.
  various patterns, 111.
  maul, 159.
  puller, 159.
  size, weight, etc., 229.
  quantities per mile, 230.
Standard cattle guard, 26.
Station grounds, 31, 45.
  platforms, 19, 28.
Steam shovel, 81.
Stewart switch, 171.
Stone, quarries, 76.
  size of, 77.
  crushers, 78.
Street-railroad crossing, 182.
Stringers, 67.
Summer work, 15.
Super-elevation, 134.
Surface cattle guards, (see Cattle guards).
Suspended and supported joints, 119.
Switches, inspection, 2.
  Wharton, 165.
  Robinson-Wharton, 166.
  split, 167.
  rods, 167.
  reenforcement, 169.
  Stewart, 169, 171.
  three-throw, 169, 170, 172, 248, 249.
  throw of, 171.
  adjustment, 171.
  slip, 178.
  leads for, 244–251.
  timbers for, 250.
Switch and lock movement, 215.
Switch lamps, 192.

INDEX.

Switch stands, 165.
  high automatic, 188–190.
  low automatic, 188–190.
  for stub switches, 190.
  jack-knife, 191.
  harp, 191.
Switch-throw adjustment, 173.

T.

Tamping bar, 158.
Tape line, 154.
Testing water, 47.
Thomson joint, 116, 117.
Ties, renewing, 15.
  tamping, 16.
  inspection, 95.
  sawed or hewed, 96.
  time for cutting, 97.
  preserving, 99.
  metal, 99.
  insulation, 100.
  spacing, tamping, etc., 102.
  weight, 229.
  quantities per mile, 230.
Tie plates, 97.
Tile drain, 59.
  shovel, 162.
Tools, 147.
  care of, 164.
Track, inspection, 3.
  walkers, 13.
  signs, 35.
  tank, 53.
  jacks, 90–94.
  bolts, 133.
  chisel, 145.
  gage, 153.
  level, 153.
  circuit, 226.
  weight of bolts, 229.
  quantities of bolts per mile, 230.
Trailing-point switches, 215.
Train signals, 197.
Trees near track, 2.
Trestle, 67.
  ballasted, 70.
  typical, 73.
  erecting, 74.

U.

Union Switch & Signal Co., switch-throw adjustment, 173.
  banner signal, 226.
Unloading rails, 127.

V.

Vaughan spring-rail frog, 186.
Velocipede cars, 147.
Verona nut lock, 111.

W.

Warren nut lock, 112.
Watchmen, 13.
Water crane, Poage, 51.
  Sheffield, 52.
  pit, 52.
Water supply, 47.
  pipe, 50.
  tank, 49.
Watson & Stillman, rail bender, 141.
  rail punch, 142.
  spike slot punch, 145.
Way freights, use of, 11.
Weber joint, 118.
  offset splice, 131.
Weeds, cutting, 15.
Weir Frog Co., rail brace, 137.
  expansion device, 139.
  reenforced switch, 172.
  switch-throw adjustment, 173.
Whitewash, 41.
Widening gage on curves, 136.
Windmills, 48.
Winter work, 14.
Wire carriers, 219.
Wire fences, 20.
  Page woven-wire, 20.
  Ellwood woven-wire, 20.
  McMullen woven-wire, 21.
  barbed wire, 21.
  expanded metal, 22.
Work trains, 11.
Wrecking force, 194.
Wrecks, duties at, 195.
Whistle signals, 199.
Wrenches, 155.

Alphabetical Index to Advertisements.

	PAGE.
Alexander Car Replacer Mfg. Co.,	12
American Steel Foundry Co.,	8
Bogue & Mills Mfg. Co.,	25
Bond Steel Post Co.,	4
Boyer & Radford,	19
Brown, M. H.,	21
Bucyrus Co., The,	24
Buda Foundry & Mfg. Co.,	13
Continuous Rail Joint Co. of America,	6
Duff Mfg. Co., The,	18
Elliot Frog & Switch Co.,	28
Ellwood Mfg. Co., The, I. L.,	9
Eureka Nut Lock Co.,	33
Eyeless Tool Co., The,	5
Fairbanks, Morse & Co.,	16, 17
Goodwin Car Co.,	23
Iron City Tool Works, Ltd.,	33
McMullen Woven Wire Fence Co., The,	10
Motley & Co., Thornton N.,	35
Norton, A. O.,	19
Oliver Iron & Steel Co.,	37
O'Neil Crossing Alarm Co., The,	26
Page Woven Wire Fence Co.,	11
Pantasote Co., The,	36
Pennsylvania Steel Co., The,	34
Poage, John N.,	30, 31
Q & C Co., The,	1, 2, 3
Roberts, Throp & Co.,	14, 15
Rodger Ballast Car Co.,	22
Tudor Iron Works,	12
Union Switch & Signal Co., The,	27
Warren Lock Washer Co.,	32
Watson-Stillman Co., The,	20
Weber Railway Joint Mfg. Co., The,	7
Weir Frog Co.,	29

CLASSIFIED INDEX TO ADVERTISEMENTS.

BALLAST CARS
  Goodwin Car Co....New York, N. Y.
  Rodger Ballast Car Co...Chicago, Ill.
BLOCK SIGNALS
  Union Switch & Signal Co., The, Swissvale, Pa.
BOLTS AND NUTS
  Motley & Co., Thornton N., New York.
  Oliver Iron & Steel Co. Pittsburgh, Pa.
  Tudor Iron Works....St. Louis, Mo.
CAR REPLACERS
  Alexander Car Replacer Mfg. Co. Scranton, Pa.
  Buda Fdy. & Mfg. Co...Harvey, Ill.
  Motley & Co., Thornton N., New York.
CATTLE GUARDS
  Fairbanks, Morse & Co. Chicago, Ill.
CROSSING ALARMS
  O'Neil Crossing Alarm Co., The, Cleveland, O.
  Union Switch & Signal Co., The, Swissvale, Pa.
CROSSING GATES
  Bogue & Mills Mfg. Co. Chicago, Ill.
CROSSINGS
  Elliot Frog & Switch Co. E. St. Louis, Ill.
  Pennsylvania Steel Co..Steelton, Pa.
  Union Switch & Signal Co., The, Swissvale, Pa.
  Weir Frog Co.........Cincinnati, O.
DUMP CARS
  Goodwin Car Co....New York, N. Y.
  Rodger Ballast Car Co..Chicago, Ill.
EXCAVATORS
  Bucyrus Co., The..S. Milwaukee, Wis.
FENCES
  Ellwood Mfg. Co., I. L...De Kalb, Ill.
  McMullen Woven Wire Fence Co., Chicago, Ill.
  Page Woven Wire Fence Co. Adrian, Mich.
FENCE POSTS
  Bond Steel Post Co....Adrian, Mich.
FROGS
  Elliot Frog & Switch Co. E. St. Louis, Ill.
  Pennsylvania Steel Co..Steelton, Pa.
  Union Switch & Signal Co., The, Swissvale, Pa.
  Weir Frog Co.........Cincinnati, O.
HAND AND INSPECTION CARS
  Buda Fdy. & Mfg. Co....Harvey, Ill.
  Fairbanks, Morse & Co. Chicago, Ill.
  Roberts, Throp & Co. Three Rivers, Mich.
INTERLOCKING SIGNALS
  Union Switch & Signal Co., The, Swissvale, Pa.
JACKS
  Boyer & Radford.........Dayton, O.
  Duff Mfg. Co., The...Allegheny, Pa.
  Fairbanks, Morse & Co. Chicago, Ill.
  Motley & Co., Thornton N., New York.
  A. O. Norton..........Boston, Mass.
  Q & C Co., The..........Chicago, Ill.
  Watson-Stillman Co., The, New York, N. Y.
NUT LOCKS
  Eureka Nut Lock Co. Pittsburgh, Pa.
  Motley & Co., Thornton N., New York.
  Oliver Iron & Steel Co. Pittsburgh, Pa.
  Warren Lock Washer Co. Boston, Mass.

PANTASOTE
  Pantasote Co., The, New York, N. Y.
PUSH CARS
  Buda Fdy. & Mfg. Co....Harvey, Ill.
  Fairbanks, Morse & Co. Chicago, Ill.
  Roberts, Throp & Co. Three Rivers, Mich.
RAIL BENDERS
  Brown, M. H........New York, N. Y.
  Buda Fdy. & Mfg. Co....Harvey, Ill.
  Watson-Stillman Co., The, New York, N. Y.
RAIL JOINTS
  American Steel Fdy. Co. St. Louis, Mo.
  Continuous Rail Joint Co. of America...........Newark, N. J.
  Tudor Iron Works....St. Louis, Mo.
  Weber Railway Joint Mfg. Co., The, New York, N. Y.
RAIL PUNCHES
  Watson-Stillman Co., The, New York, N. Y.
RAIL SAWS
  Q & C Co., The..........Chicago, Ill.
RAILS
  Pennsylvania Steel Co..Steelton, Pa.
SIGNAL POSTS
  Bond Steel Post Co....Adrian, Mich.
SPIKES
  Motley & Co., Thornton N., New York.
  Tudor Iron Works....St. Louis, Mo.
SWITCHES
  Elliot Frog & Switch Co. E. St. Louis, Ill.
  Pennsylvania Steel Co..Steelton, Pa.
  Union Switch & Signal Co., The, Swissvale, Pa.
  Weir Frog Co.........Cincinnati, O.
SWITCH STANDS
  Buda Fdy. & Mfg. Co....Harvey, Ill.
  Elliot Frog & Switch Co. E. St. Louis, Ill.
  Pennsylvania Steel Co..Steelton, Pa.
  Union Switch & Signal Co., The, Swissvale, Pa.
  Weir Frog Co.........Cincinnati, O.
TANK VALVES
  Fairbanks, Morse & Co. Chicago, Ill.
  Poage, John N........Cincinnati, O.
TIE PLATES
  Motley & Co., Thornton N., New York.
  Q & C Co.,............Chicago, Ill.
TRACK TOOLS
  Buda Fdy. & Mfg. Co....Harvey, Ill.
  Eyeless Tool Co., The, New York, N. Y.
  Fairbanks, Morse & Co. Chicago, Ill.
  Iron City Tool Works, Ltd. Pittsburgh, Pa.
  Motley & Co., Thornton N., New York.
  Oliver Iron & Steel Co. Pittsburgh, Pa.
  Q & C Co., The..........Chicago, Ill.
WATER COLUMNS AND TANKS
  Fairbanks, Morse & Co. Chicago, Ill.
  Poage, John N........Cincinnati, O.
  Motley & Co., Thornton N., New York.

~ THE ~
SERVIS TIE PLATE

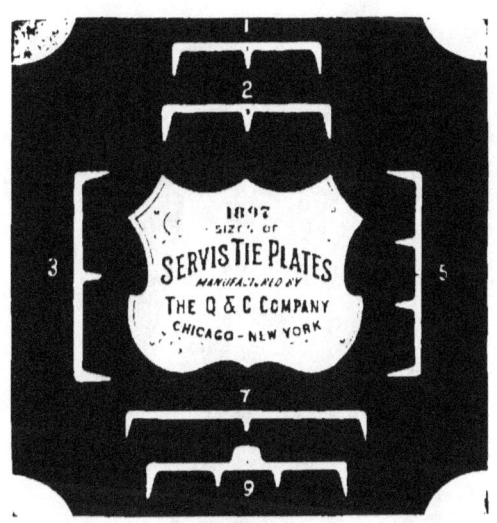

Only Tie Plate in successful use for a period of ten years

THE Q & C Co.

CHICAGO   NEW YORK

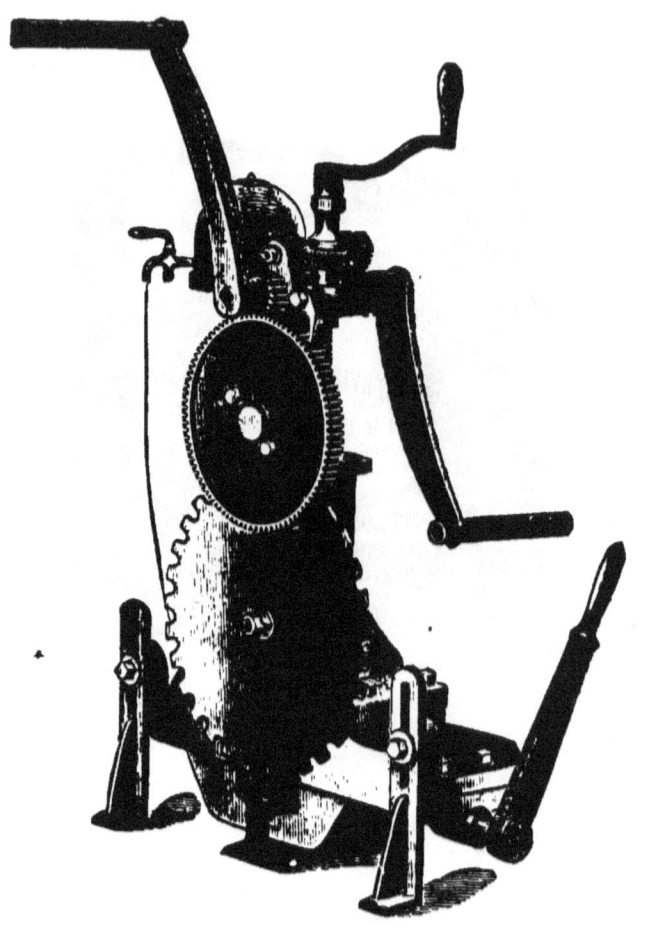

THE BRYANT RAIL SAW

SAVES TIME AND MONEY

THE Q & C Co.,

CHICAGO   NEW YORK

"The Best is as Good as any"

The Q & C Compound Lever Jacks

MADE IN 19 SIZES

For All Purposes

THE Q & C SELF-FEEDING RAIL DRILL

Furnished with

    OVER OR
    UNDER-RAIL
    CLAMPS

Will Drill a $\frac{7}{8}$-inch Hole in Less than One Minute

Q & C COMPANY

    CHICAGO
    NEW YORK

STEEL BRIDGES
ARE
UNSAFE

IF MADE OF LIGHT INFERIOR STOCK.

BOND STEEL POSTS

FOR

R. R. Signals, Fences, &c.,

are made on the

SAFE BRIDGE PRINCIPLE.

HEAVY PLATES
    BEST QUALITY STEEL
        SCIENTIFICALLY PLANNED
            HONESTLY MADE

They are covered inside and out with a coating which is not affected by air or water, acids or alkali. Used by United States Government, and many first-class railroads. Write for Circulars.

BOND STEEL POST CO.
ADRIAN, MICH.

THE EYELESS TOOL CO.

26 Cortlandt St.

NEW YORK, - U. S. A.

MANUFACTURERS OF

Eyeless Steel Picks

Standard Railroad Track Tools

Machinists', Blacksmiths'

and Mining Tools

OUR PRODUCT IS

Guaranteed to be Unsurpassed

in material, workmanship and general finish, and is

STANDARD

on many of the largest

RAILROAD SYSTEMS

in the United States.

TOOLS OF SPECIAL DESIGN

made to Railroad Companies' drawings.

Continuous Rail Joint Company of America

SOLE MANUFACTURERS

912 Prudential Building,   Newark, N. J.

Patented in United States and Europe

Rail Joints, Step Joints and Insulating Joints
ALL OF THE CONTINUOUS PATENT TYPE

Rapidly Taking the Place of Angle Bars
In Successful Use on 78 Railroads

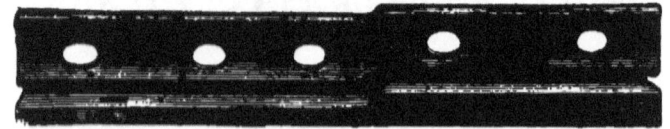

Connecting 85-Pound and 70-Pound Rail with the Continuous Step Joint

ROBERT GRAY, Jr., *President*   L. F. BRAINE, *General Manager*
F. C. RUNYON, *Secretary*   F. T. FEAREY, *Treasurer*

THE WEBER RAILWAY JOINT MFG. CO.

Standard Joint

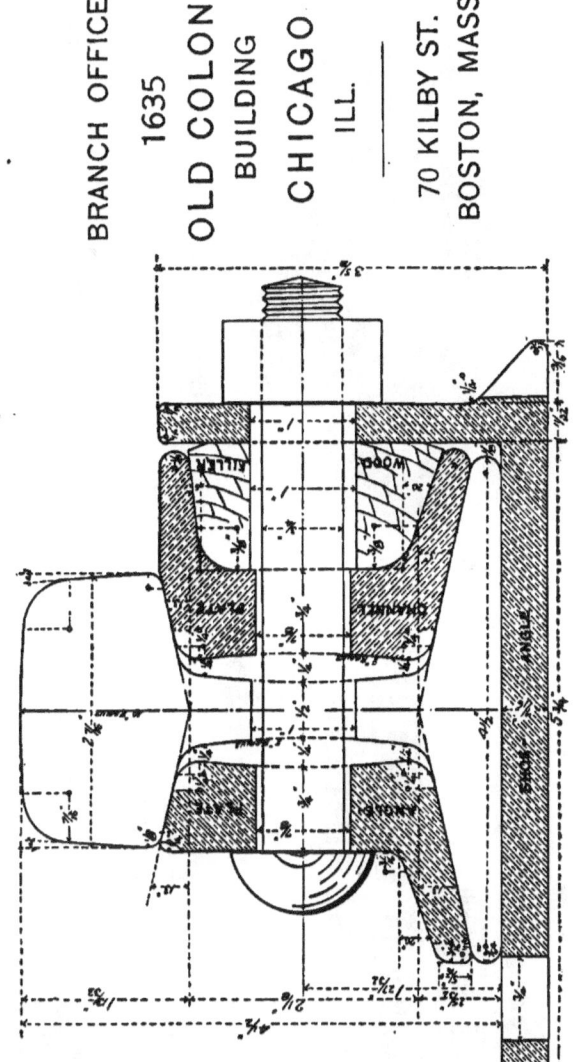

BRANCH OFFICES

1635 OLD COLONY BUILDING
CHICAGO, ILL.

70 KILBY ST.
BOSTON, MASS.

COTTON EXCHANGE BUILDING
NEW YORK, N.Y.

Our Specialty ~
BASIC STEEL RAILROAD CASTINGS
OF EVERY DESCRIPTION,

FOR TRACK, BRIDGES, CARS AND ENGINES.

AMERICAN STEEL FREIGHT TRUCK.
CATALOGUE ON APPLICATION.

AMERICAN STEEL FOUNDRY CO.
ST. LOUIS, MO. ~

SEND FOR COMPLETE CATALOGUE.

AMERICAN SUSPENSION RAIL CLAMP.
SEND FOR BOOKLET.

Established 1873

THE I. L. ELLWOOD MFG. CO.
DE KALB, ILLINOIS
MANUFACTURERS AND CONTRACTORS OF

Steel Wire Railroad Fences

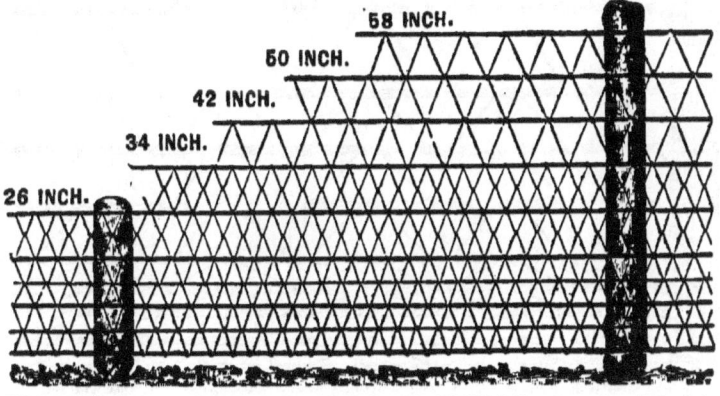

ELLWOOD WOVEN RAILROAD FENCE (Standard Style)

WE GUARANTEE

GREATEST DURABILITY

GREATEST EFFICIENCY

GREATEST ECONOMY

Estimates promptly furnished and contracts taken in any part of the country for RAILROAD FENCES, either Ellwood Woven Railroad fence, genuine Glidden Barb Wire, or the two combined.

Can furnish **everything** or such part of the work and material only as may be desired. Having several large Railway Fencing Outfits, our facilities for doing this work quickly, satisfactorily and at the least possible cost are unequaled. All work and material fully guaranteed.

43 PAGE CATALOGUE SENT ON APPLICATION.

McMullen's Railroad Fencing

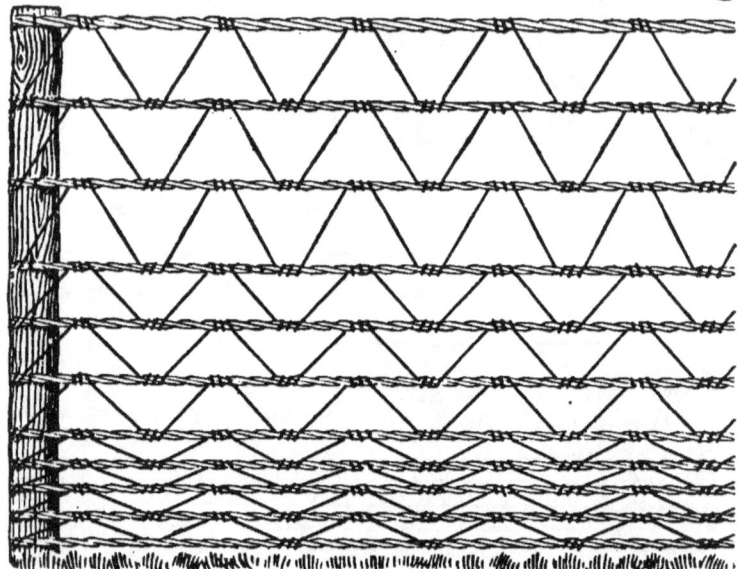

The above cut represents our Steel Wire Cable Fencing

The above cut represents our Spring Steel Wire Fencing with horizontal wires spirally curved

We make both of the above kinds of fencing in all the various widths, and with narrow meshes at bottom, as desired. We can also furnish them in especially heavy wire if desired. Correspondence solicited.

THE McMULLEN WOVEN WIRE FENCE Co. — CHICAGO

A GOOD TRACK, well ballasted and in perfect alignment, is the pride and joy of the progressive roadmaster, and a FIRST-CLASS FENCE enclosing right-of-way—one that fills requirements and preserves its good appearance, is his comfort and his peacemaker.

PAGE WOVEN WIRE FENCE CO.,
ADRIAN, MICH.

TUDOR IRON WORKS

ST. LOUIS, - MO.

MANUFACTURERS OF

Track Spikes, Bolts and Splices

Boat Spikes, Bridge Bolts and Blank Nuts

Bar Iron and Steel

PRESSED STEEL CAR AND ENGINE REPLACER

THE BEST IN THE WORLD

GUARANTEED TO CARRY 100 TONS

SEND FOR CIRCULAR

ALEXANDER CAR REPLACER MFG. CO.

SCRANTON, PA.

BUDA FOUNDRY & MFG. CO.

MANUFACTURERS OF

Standard Section Hand & Push Cars

Railway Velocipedes

Switch Stands and Fixtures

Paulus Automatic-fed Track Drills

Track Gauges, Car Replacers

Rail Benders, etc., etc.

BUDA FOUNDRY & MFG. CO.

CHICAGO OFFICE
917 MONADNOCK BLOCK

WORKS AT
HARVEY, ILL

THE CYRUS ROBERTS
IMPROVED HAND CARS

EQUIPPED WITH improved Gallows Frame, which can be detached and removed by simply loosening the Main Clamping Bolt and disconnecting Pitman Turnbuckle.

THE DRIVING MECHANISM can be thrown in or out of gear by means of our Patented Slip Pinion.

THE ENTIRE GEARING is located beneath the Platform, so that in removing the Gallows Frame the entire platform surface can be utilized for the uses of a regular Push Car; has proven especially serviceable for track patrol as an emergency Car.

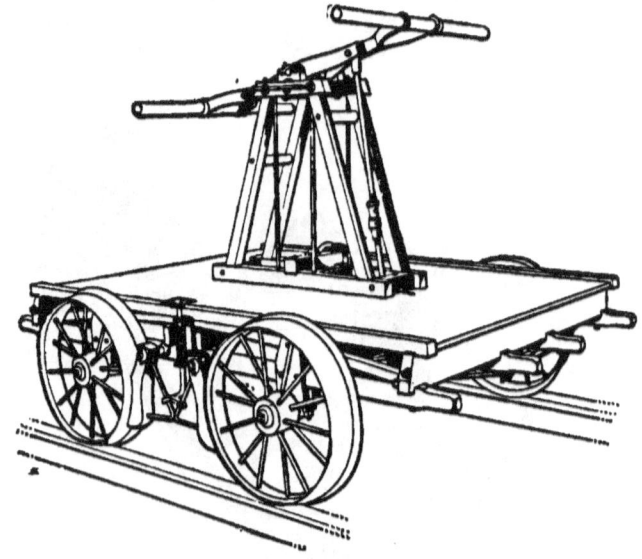

No. 1 Car.

IMPROVED BRAKE, giving much greater leverage than the ordinary pattern of brake.

THE DIAGONAL TRUSS RODS connected to each corner of the car, by means of turnbuckle the car body can be kept squared properly, boxes kept in line, and all twisting or "wringing" tendency of car frames, as found in all other types of light cars, is entirely overcome.

THE STEEL PEDESTALS provide an unmovable seat for axle boxes and journals, reducing friction.

ROLLER BEARINGS if desired.

THE RESULTS FROM ACTUAL TESTS show a great saving in power required to propel our Car as compared with the ordinary type of Walking Beam Car.

ROBERTS, THROP & CO.
THREE RIVERS, MICH.

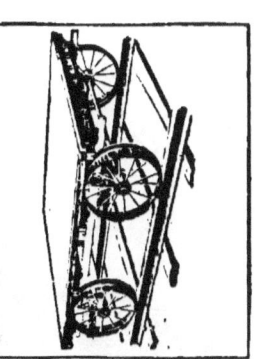

"CYRUS ROBERTS"

Patented Hand Cars

SEND FOR
CATALOGUE GIVING
OFFICIAL TESTS

ROBERTS, THROP & CO.

THREE RIVERS, MICH.

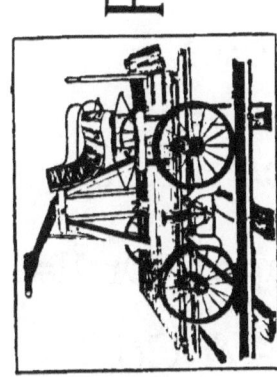

ALL KINDS
OF
LIGHT CARS

SECTION HAND CARS

LIGHTEST EASIEST RUNNING
MOST DURABLE

Pressed Steel Wheels, Machine Cut Gears, Tapering Wheel Fit, no Keys required, Largest variety of designs for all purposes

PUSH CARS, TRACKLAYING CARS, VELOCIPEDES, GASOLINE MOTOR CARS, CATTLE GUARDS

Track Levels, Gauges, Drills

FAIRBANKS, MORSE & CO.

JACKS

AUTOMATIC LOWERING JACK

"TRIP" JACK

For Track Work
Ballasting   Bridge Gangs
Car Repairing, Oil Boxes, &c.

LARGE NUMBER OF MODIFICATIONS

GASOLINE PUMPING ENGINES
  IMPROVED STEAM PUMPS
    IMPROVED WATER COLUMNS
      TANKS AND FIXTURES
        WATER STATIONS ERECTED

FAIRBANKS, MORSE & CO.
CHICAGO

Cleveland	Louisville	Kansas City	San Francisco
Cincinnati	St. Louis	Omaha	Los Angeles
Indianapolis	St. Paul	Denver	Portland, Ore.

THE BARRETT TRACK JACK

Recommended as a Standard by the
ROAD MASTERS' ASSOCIATION OF AMERICA

These Jacks are made with Malleable Iron Frames, Steel Pawls, Machinery Steel Bearings and Pivots, Forged Steel Rack. The wearing parts are removable and readily renewable at slight expense. The rectangular base gives great lifting strength and fits into close quarters better than other shaped stands. Adaptable to either high or low set loads.

The Barrett Jack is the safest, best and strongest known to the railroad world to-day.

Lifting capacity, 10 to 15 tons.

No. 2, Automatic Lowering

No. 1, Trip

FOR CATALOGUE
AND PRICE LIST
APPLY TO

THE DUFF MANUFACTURING CO.

Marion and Martin Avenues

ALLEGHENY, PA.

Quick Drop Track Jack

No. 18 A   Maxon Pat.

QUICK AND POSITIVE, SAFE AND DURABLE

SIMPLE IN CONSTRUCTION

Load can be raised or lowered one or two notches at a time or dropped instantly.

Ratchet bars reinforced a full length by a ¾ inch iron bolt, which gives more rigidity and strength. Impossible to throw the track out of line.

Hardened steel bushings and pins used.

SIZE OF BASE 7 X 12 INCHES.

SEND FOR CATALOGUE AND DISCOUNTS.

MANUFACTURED BY

BOYER & BRADFORD
DAYTON, O.

THE GREATEST VALUE EVER OFFERED
IN A *TRACK JACK*
TO THE ROADMASTERS OF AMERICA

Gentlemen:—Are you looking for a good SAFE Track Jack, at a PRICE TO SUIT THE TIMES? How would something of this kind strike you for all round work?

Say a Jack 24 inches high, to weigh 60 lbs.—light enough for section work.

Capacity 10 tons—powerful enough for ordinary yard work.

Raise of bar, 15 inches clear—enough for ordinary ballasting or new work, one Jack taking the place of *two*.

Easy and positive trip—cannot be *Stuck* under any conditions.

With only Six (6) Pieces and Two (2) Pins in the whole Jack.

Thoroughly made of *Malleable Iron and Steel* throughout, wearing parts all *Hardened Steel Interchangeable.*

HERE IT IS
THE "NORTON" "SURE DROP," No. 5

THE NEAREST PERFECTION of any Jack made up to date

WE MAKE IT, YOU TAKE IT and TRY IT; IF YOU LIKE IT, BUY IT at $14.25 LIST.

This Jack, after trial in competition with all the leading Jacks, has been adopted as standard by the Canadian Pacific Ry. Co. on its system.

A. O. NORTON, 167 Oliver St., BOSTON, MASS.

The Watson-Stillman Co.
204–210 E. 43d St., New York

HYDRAULIC JACKS

HYDRAULIC Rail Benders

RAIL PUNCHES

Full Line of Hydraulic Tools and Machinery

The Emerson Patent Rail Bending and Straightening Machine

FOR TEE AND STREET RAILS

⟨ TRADE MARK EMERSON ⟩

MADE OF WROUGHT IRON AND WROUGHT STEEL

Bends, Straightens and Takes Kinks out of Rails already laid

PERFECT for TAKING OUT SURFACE BENDS

Immediate Shipments.

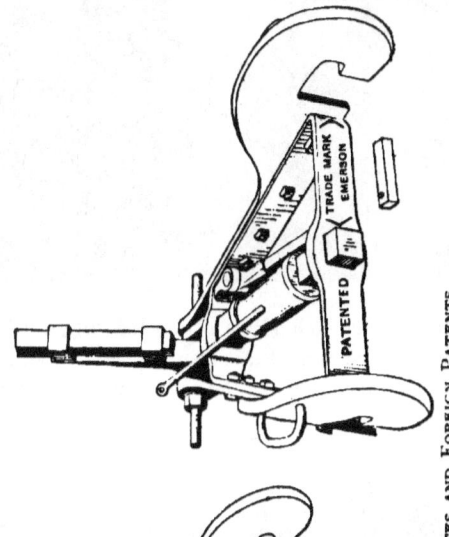

United States and Foreign Patents.

HIGHEST TESTIMONIALS FURNISHED. CORRESPONDENCE SOLICITED.

M. H. BROWN,

96 BROADWAY, N. Y.

Sole Owner and Manufacturer

Size No. 1 for Tee Rails up to 45 lbs. to yard.
" No. 2 " " " 45 " 65 " " "
" No. 3 " " " 65 " 90 " " "
" No. 4 " " " 90 " 115 " " "

Also No. 4 machine fitted to order with steel dies for bending street rails of ordinary or peculiar shape.

RODGER BALLAST CARS

Showing track after unloading a train load of gravel. The distributing car was left off the rear of this train while unloading in order to show the ridge of ballast left between the rails while unloading.

Showing condition in which same track was left after the distributing car had been run over it. The distributing car is always at rear of train while unloading and always leaves track **cleared** and flanged as shown.

For Illustrated Catalogue and further information address

RODGER BALLAST CAR CO.
1301 FISHER BUILDING

GOODWIN CAR CO.

OWNING AND OPERATING GOODWIN PATENTS

CARS MANUFACTURED AND LEASED

CHICAGO
115 Dearborn Street

NEW YORK
96 Fifth Avenue

These cars will carry and discharge Pig Iron, Grain, Ore, Coal, Gravel, Clay or

ANY MATERIAL
IN
ANY CONDITION
DRY, WET, OR FROZEN

They dump on either or both sides or in the centre, the load being discharged without the use of crow-bar, pick, mallet or shovel.

No other car can dump a frozen load.

These Cars ballast on both sides of the track or between the rails. No plow or HAND LABOR required to free the track.

They ballast on the run, distribute their load evenly, and the quantity of discharge can be regulated either by rate of speed or by the operator.

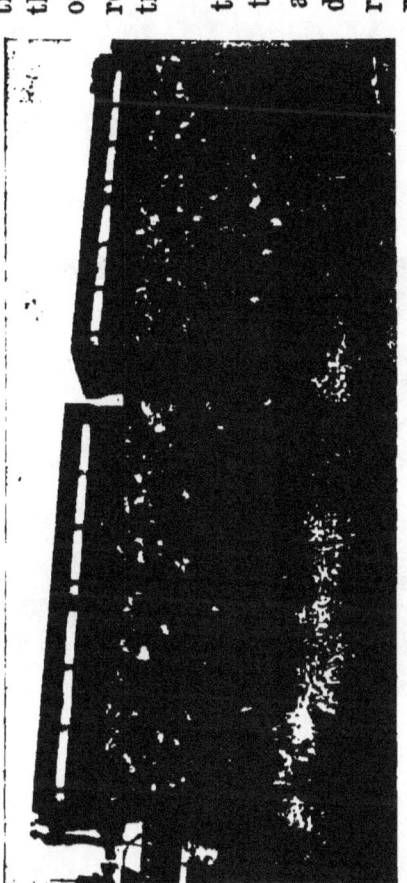

ALL STEEL GOODWIN GRAVITY DUMPING CARS
FILLING IN TRESTLE WITH RUBBLE STONE

CAPACITY 40,000 LBS. TO 125,000 LBS.

ADDRESS CORRESPONDENCE TO NEW YORK OFFICE

THE BUCYRUS COMPANY
SO. MILWAUKEE, WIS.

BUILDERS OF

DREDGING AND EXCAVATING MACHINERY

FOR EVERY SERVICE

No.	WEIGHT	SIZE DIPPER
1	12 tons	½ yards
2	20 "	1 "
3	35 "	1½ "
4	45 "	1¾ "
5	55 "	2 "
6	65 "	2½ "
7	75 "	3 "

FOR FURTHER INFORMATION ADDRESS

THE BUCYRUS COMPANY
SO. MILWAUKEE, WIS.

BOGUE & MILLS MFG. CO., 218 La Salle St., Chicago, Ill., U.S.A.

MANUFACTURERS OF

Street and Crossing Guards, being Pneumatic Lever and Cable Gates for protection of street crossings of Railways at grade and Draw-Bridges.

Of the roads using our gates we refer you to the following:

Baltimore & Ohio.
Chicago & Alton.
Chicago, Burl'n & Quincy.
Chicago & Eastern Illinois.
Chicago & Grand Trunk.
Chicago & Northern Pacific.
Chicago & North-Western.
Chicago & Western Indiana.
Chicago & West Michigan.
Clev'd, Cin., Ch'go & St. l.
Delaware & Hudson Canal Co.
Denver & Rio Grande.
Michigan Central.
Erie Railroad.
Evansville & Terre Haute.
Great Northern R'y Co.
Illinois Central.
Indiana, Illinois & Iowa.
K. City, Fort Scott & Memp.
New York, Ont. & Western.
Kansas City Belt Line.
Lehigh Valley.
Long Island Railroad.
L'ville New Alb'y & C'go.
Louisville, New Orleans & Texas.
Northern Pacific.
Oregon Imp. Co
Pitt'g, Ft Wayne & C'go.
Southern Pacific Co.
St. L. San Francisco.
Pittsburg, Cin., Chicago & St Louis.
Pittsburg & Lake Erie.
Toledo & Ohio Central.
Texas Pacific.
Union Pacific.
Wisconsin Central.

BOGUE & MILLS No. 1

BOGUE & MILLS No. 2, STANDARD

THE O'NEIL
HIGHWAY CROSSING ALARM

EIGHT YEARS SUCCESSFUL USE

on 40 of the trunk line railroads has proven the O'NEIL HIGHWAY CROSSING ALARM to be the cheapest and most economical and reliable safeguard known for the protection of public highway crossings.

Their use on all dangerous and obscure highway crossings would save many valuable lives and prove of great economy to all railroads using them.

They are always on guard and sure to sound the alarm on approach of train.

WRITE FOR FURTHER INFORMATION.

THE O'NEIL CROSSING ALARM CO.
CLEVELAND, OHIO

SWITCH AND SIGNAL COMPANY

OF SWISSVALE, ALLEGHENY COUNTY, PA.

U. S. A.

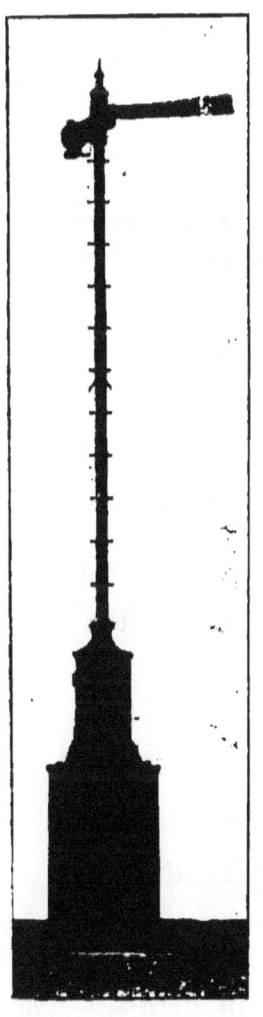

Designers and Manufacturers of Interlocking and Block Signal Appliances, Frogs, Slips, Switches, Switch Stands, etc.

Sole Manufacturers and Patent Owners of the Westinghouse Electro Pneumatic Interlocking and Block Signals; Automatic Track Circuit Block Signals of the Disc, Banner or Semaphore patterns; Electric Locking; The Union Block and Lock System; The Baker-Knight Lock and Block System; Electric Crossing Alarm Bells; Special Appliances for the protection of Draw-Bridges, Tunnels, etc.

In fact, everything in the Signaling or Interlocking line.

Plans and Estimates furnished on application. All Materials and Workmanship guaranteed.

Elliot Frog & Switch Co.

EAST ST. LOUIS, ILLINOIS

MANUFACTURERS OF

Rigid and Spring Rail Frogs, Split Switches of New and Improved Designs, Switch Stands, Stub Switch Fixtures, etc.

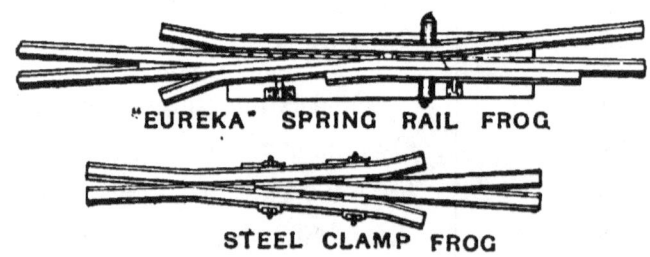

"EUREKA" SPRING RAIL FROG

STEEL CLAMP FROG

SLIP SWITCH CROSSINGS

Switches and Movable Points operated successfully by one HASTY-ELLIOT Switch Stand.

THREE THROW SPLIT SWITCHES

operated by same Stand. It is simple and with few pieces. We make the Stand High or Low Pattern.

Make any other special work in Frogs, Crossings and Switches from plans furnished.

CATALOGUE AND ESTIMATES FURNISHED ON APPLICATION

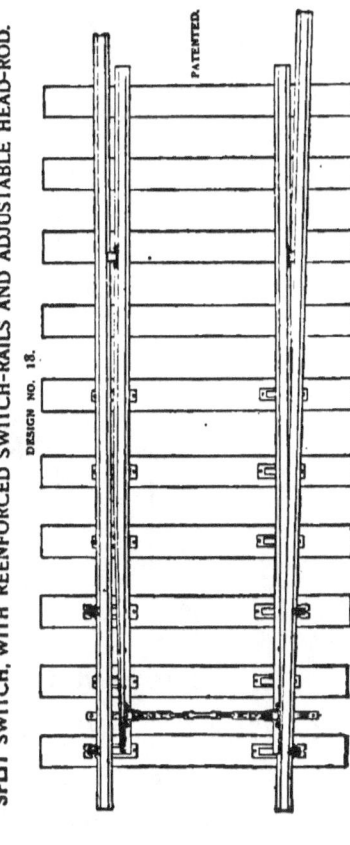

WEIR FROG CO. Cincinnati, Ohio

MODERN TRACK EQUIPMENT

ADJUSTABLE REINFORCED SPLIT SWITCHES
AUTOMATIC "CLUTCHLESS" SPRING SWITCH STANDS
IMPROVED SPRING FROGS   RAIL BRACES
RIGID FROGS: BOLTED CLAMPED and PLATE PATTERNS

SPLIT SWITCH, WITH REENFORCED SWITCH-RAILS AND ADJUSTABLE HEAD-ROD.

DESIGN NO. 18.

SPRING-RAIL FROG.
DESIGN NO. 19.

WATER COLUMN
FOR TAKING WATER FROM WATER WORKS or TANKS

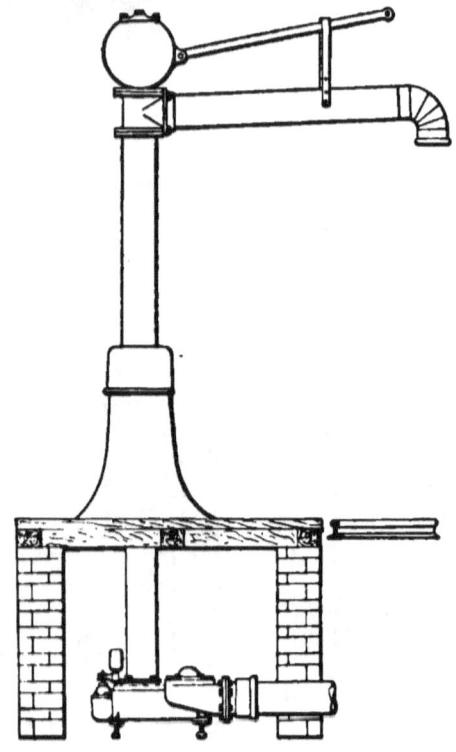

DISTINCTIVE FEATURES:

IT CLOSES ITS OWN VALVE WITHOUT CONCUSSION
ONE MAN OPERATES IT
IT WILL NOT FREEZE UP
IT IS AUTOMATIC IN ITS MOVEMENTS

JOHN N. POAGE
SOLE MANUFACTURER
CINCINNATI, OHIO

POAGE'S TANK VALVES

WITH UNIVERSAL JOINT

CONNECTED WITH BOTTOM OR SIDE OF OPEN OR ENCLOSED TANKS

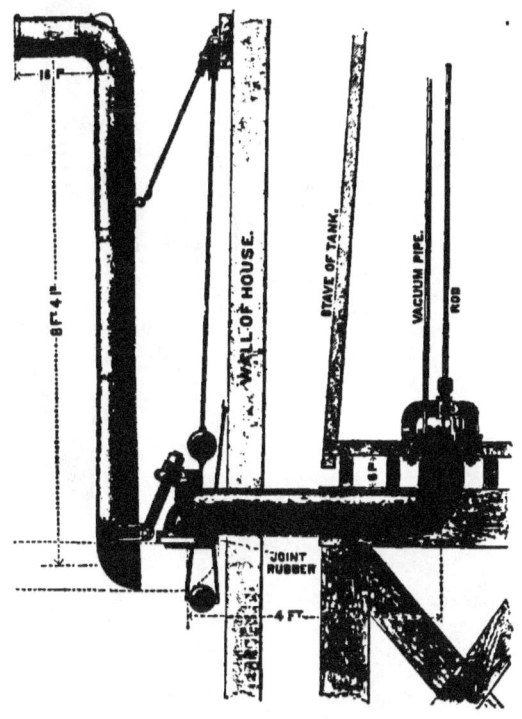

Is raised and lowered the same as the common tank valve.

It may be moved around laterally.

It has a perfect universal joint.

This construction obviates the necessity of bringing the train exactly opposite the spout.

JOHN N. POAGE

SOLE MANUFACTURER

CINCINNATI, OHIO

IF YOU WANT ABSOLUTE SAFETY AND ECONOMY COMBINED, USE THE

WARREN LOCK WASHERS

Six years in use by the railroads alone is sufficient proof of their great merit; besides the numerous other consumers who have used large quantities. Workmanship first class. Made of fine steel. Every Washer warranted. Tons in use.

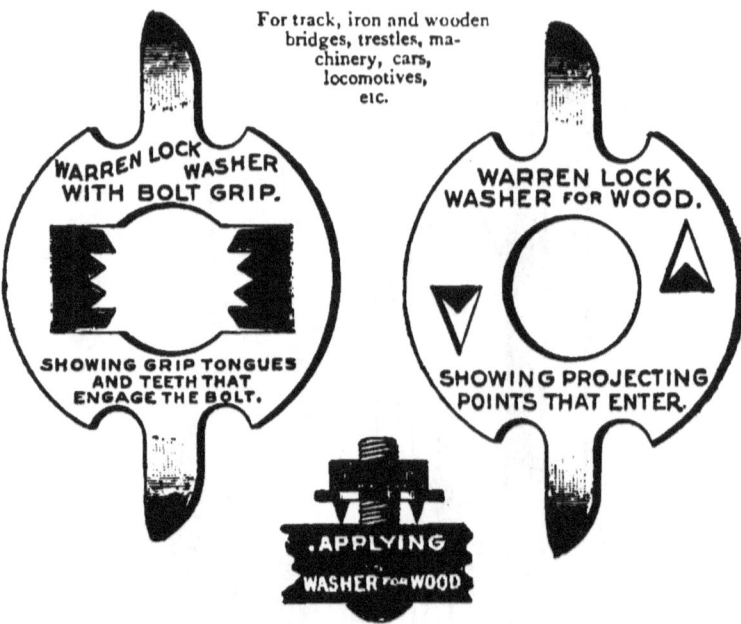

For track, iron and wooden bridges, trestles, machinery, cars, locomotives, etc.

The only general utility and positive lock washer that can be relied on. Saves 25 per cent. in common washers, check nuts and extra length of bolt. Can be used many times. Easy to apply and remove.

PRICES RIGHT   SAMPLES FREE   CORRESPONDENCE SOLICITED

Refer to: New York Central, Boston & Maine, Boston & Albany, Concord & Montreal, Boston & Lowell, and many others

SPECIAL INDUCEMENTS TO RAILROADS

WARREN LOCK WASHER CO.

CHAS. H. WARREN, Mgr.

Washington and Norfolk Sts., Dorchester District, **BOSTON, MASS.**

PITTSBURGH

Makers of Standard R. R. Track Tools

THE BEST TRACK TOOLS

THE BEST NUT LOCK

THE STRONGEST SPRING LOCK EVER MADE

Far exceeds any device ever used for holding nuts and bolts secure on track joints, cars, car trucks, bridges, engines, etc.

Eureka Nut Lock Company

THE PENNSYLVANIA STEEL CO.
STEELTON, PENNA.

STEEL RAILS All PATTERNS & SIZES for Steam & Elec. Roads

BRIDGES AND BUILDINGS

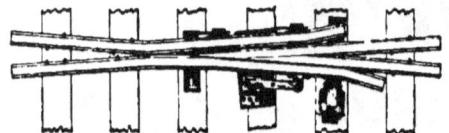

FROGS AND CROSSINGS

FOR ALL KINDS OF TRACKS. Special designs of extreme durability and safety. Several other designs of less cost: also several kinds of SLIPS (combining Crossing and Switches) and fixtures for same.

SPLIT OR POINT SWITCHES

with the usual number of connecting bars of various patterns now "Standard" on many roads. Some with parts of SOLID IRON, machine forged, that are elsewhere welded. Some with point rails variously reenforced; also our Special Switch (see sectional view) with point rails STIFFENED by special angle, requiring only one connecting bar, adjustable as shown, or plain.

THREE-THROW (DOUBLE THROW) SPLIT SWITCHES
of various patterns. Valuable for Three Way Turnouts.

IMPROVED SWITCH STANDS

This Low Target Stand is an especial favorite for use in yard tracks, where the low down weighted lever, parallel to track, easy to handle, holding switch secure, makes it extremely convenient. Is also widely used for main tracks. Another favorite is the LONG SAFETY STAND, which lets train through trailing, but holds switch to the position, as locked, for either track, and detects any obstruction in switch. Various other patterns of Upright Stands, Ground Levers, etc.

GUARD RAIL CLAMP

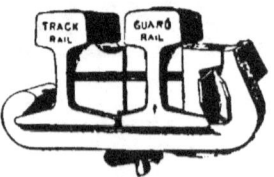

Secures the Guard Rail immovably on main track rail. Variable flangeway, for various gauges or to compensate for wear of guard rail. Applied without drilling of main track rail.

WROUGHT IRON STUB SWITCH TIE BARS, for all sizes of Rail.

RAIL BRACES of Pressed Steel. **RAIL BENDERS** with Racket Wrench.

www.ingramcontent.com/pod-product-compliance
Lightning Source LLC
Chambersburg PA
CBHW021956220426
43663CB00007B/838